Repairing and Extending Finishes

VAN NOSTRAND REINHOLD'S
BUILDING RENOVATION AND RESTORATION SERIES

Repairing and Extending Weather Barriers   ISBN 0-442-20611-9

Repairing and Extending Finishes, Part I   ISBN 0-442-20612-7

Repairing and Extending Finishes, Part II   ISBN 0-442-20613-5

Repairing and Extending Nonstructural Metals   ISBN 0-442-20615-1

Repairing and Extending Doors and Windows   ISBN 0-442-20618-6

Repairing, Extending, and Cleaning Brick and Block   ISBN 0-442-20619-4

Repairing, Extending, and Cleaning Stone   ISBN 0-442-20620-8

Repairing and Extending Wood   ISBN 0-442-20621-6

Repairing and Extending Finishes

PART II

ACOUSTICAL TREATMENT

RESILIENT FLOORING

PAINT

TRANSPARENT FINISHES

H. Leslie Simmons, AIA, CSI

VNR VAN NOSTRAND REINHOLD
New York

Library of Congress Catalog Card Number 88-36478
ISBN 0-442-20613-5

Printed in the United States of America

Designed by Caliber Design

Van Nostrand Reinhold
115 Fifth Avenue
New York, New York 10003

Van Nostrand Reinhold International Company Limited
11 New Fetter Lane
London EC4P 4EE, England

Van Nostrand Reinhold
480 La Trobe Street
Melbourne, Victoria 3000, Australia

Nelson Canada
1120 Birchmount Road
Scarborough, Ontario M1K 5G4, Canada

16 15 14 13 12 11 10 9 8 7 6 5 4 3 2 1

Library of Congress Cataloging-in-Publication Data
(Revised for vol. 3)

Simmons, H. Leslie.
  Repairing and extending finishes.

  (Building renovation and restoration series)
  Includes bibliographical references.
  Contents: pt. 1. Plaster, gypsum board, and ceramic tile—pt. 2. Acoustical treatment, resilient flooring, paint and transparent finishes
  1. Buildings—Repair and reconstruction. 2. Finishes and finishing. I. Title.
TH3411.S55 1990   698'.028'8   88-36478
  ISBN 0-442-20612-7 (v. 2)
  ISBN 0-442-20613-5 (v. 3)

Contents

Series Foreword   ix
Preface   xiii
Acknowledgments   xix

CHAPTER 1 Introduction   1

What Finishes Are and What They Do   2
Failure Types and Conditions   2
What to Do in an Emergency   4
Professional Help   4
  Help for Building Owners   4
  Help for Architects and Engineers   11
  Help for General Building Contractors   12
Prework On-site Examination   13
  The Owner   13
  Architects and Engineers   13
  Building Contractors   14

CHAPTER 2 Structure and Substrates   16

Structural Framing Systems   17
  Steel and Concrete Structure Failure   18
  Wood Structure Failure   18
  Structure Movement   23
Substrates   24
  Solid Substrate Problems   24
  Supported Substrate Problems   26
Other Building Elements   29
  Other Building Element Problems   29
Where to Get More Information   31

CHAPTER 3 Acoustical Treatment   33

Acoustical Ceilings   34
  Acoustical Materials   35
  Support Systems   39
  Accessories and Miscellaneous Materials   50
  Fire Performance   51
  Installation   51
  Why Acoustical Ceilings Fail   58
  Evidence of Failure   66
  Cleaning, Repairing, Refinishing, and Extending Acoustical Ceilings   70
  Installing New Acoustical Ceilings over Existing Materials   85
Integrated Ceilings   89
  System Components   89
  Installation   91
  Why Integrated Ceilings Fail   92
  Evidence of Failure   92
  Cleaning, Repairing, Refinishing, and Extending Integrated Ceilings   93
  Installing New Integrated Ceilings over Existing Materials   93
Acoustical Wall Panels and Baffles   93
  Panel and Baffle Materials and Fabrication   94
  Support Systems   100
  Fire Performance   103
  Installation   105
  Why Acoustical Wall Panels and Baffles Fail   109
  Evidence of Failure   114
  Cleaning, Repairing, Refinishing, and Extending Acoustical Wall
    Panels and Baffles   117
  Installing New Acoustical Wall Panels and Baffles over
    Existing Materials   128
Where to Get More Information   130

CHAPTER 4   Resilient Flooring   132

Resilient Flooring Materials   **133**
   Characteristics Common to All Types of Resilient Flooring   133
   Tile   136
   Sheet Flooring   140
   Miscellaneous Resilient Units and Accessories   143
   Underlayment, Primers, and Adhesives   147
   Cleaning and Polishing Materials   149
Resilient Flooring Installation   **149**
   Requirements Common to All Resilient Flooring   149
   Installing Resilient Tiles   154
   Installing Resilient Sheet Flooring   155
   Installing Miscellaneous Resilient Units and Accessories   156
   Cleaning, Protection, and Finishing of Resilient Materials   159
Resilient Flooring Failures and What to Do about Them   **160**
   Why Resilient Flooring Fails   160
   Evidence of Failure   164
Cleaning, Repairing, and Refinishing Resilient Flooring   **167**
   General Requirements   167
   Cleaning and Refinishing Resilient Materials   169
   Repairing Resilient Materials   170
Installing New Resilient Flooring over Existing Materials   **175**
   Requirements Common to All Resilient Flooring   175
   Installing Tiles, Sheet Flooring, Other Units, Accessories, and
      Miscellaneous Materials   183
   Cleaning, Protection, and Finishing   183
Where to Get More Information   **183**

CHAPTER 5   Paint and Transparent Finishes   185

Paint and Transparent Finish Materials   **187**
   Characteristics Common to All Paint and Transparent
      Finish Materials   187
   Paints   190
   Transparent Finishes   197
   Miscellaneous Materials   199
Paint and Transparent Finish Applications   **200**
   Requirements Common to All Paint and Transparent
      Finish Applications   200
   Preparation for Painting and Applying Transparent Finishes   203
   Applying Paint and Transparent Finishes   213
Paint and Transparent Finish Failures and What to Do about Them   **216**
   Why Paint and Transparent Finishes Fail   216

Evidence of Failure   229
Cleaning and Repairing Paint and Transparent Finishes   237
Applying Paint and Transparent Finishes over Existing Surfaces   **247**
Requirements Common to All Paint and Transparent Finishes   247
Preparation for Painting and Applying Transparent Finishes   251
Paint and Transparent Finish Applications   263
Where to Get More Information   **267**
Appendix: Data Sources   277
Bibliography   283
Index   297

Series Foreword

To spite a national trend toward renovation, restoration, and remodeling, construction products producers and their associations are not universally eager to publish recommendations for repairing or extending existing materials. There are two major reasons. First, there are several possible applications of most building materials; and there is an even larger number of different problems that can occur after products are installed in a building. Thus, it is difficult to produce recommendations that cover every eventuality.

Second, it is not always in a building construction product producer's best interest to publish data that will help building owners repair their product. Producers, whose income derives from selling new products, do not necessarily applaud when their associations spend their money telling architects and building owners how to avoid buying their products.

Finally, in the *Building Renovation and Restoration Series* we have a reference that recognizes that problems frequently occur with materials used in building projects. In this book and in the other books in this series,

Simmons goes beyond the promotional hyperbole found in most product literature and explains how to identify common problems. He then offers informed "inside" recommendations on how to deal with each of the problems. Each chapter covers certain materials, or family of materials, in a way that can be understood by building owners and managers, as well as construction and design professionals.

Most people involved in designing, financing, constructing, owning, managing, and maintaining today's "high tech" buildings have limited knowledge on how all of the many materials go together to form the building and how they should look and perform. Everyone relies on specialists, who may have varying degrees of expertise, for building and installing the many individual components that make up completed buildings. Problems frequently arise when components and materials are not installed properly and often occur when substrate or supporting materials are not installed correctly.

When problems occur, even the specialists may not know why they are happening. Or they may not be willing to admit responsibility for problems. Such problems can stem from improper designer selection, defective or substandard installation, lack of understanding on use, or incorrect maintenance procedures. Armed with necessary "inside" information, one can identify causes of problems and make assessments of their extent. Only after the causes are identified can one determine how to correct the problem.

Up until now that "inside" information generally was not available to those faced with these problems. In this book, materials are described according to types and uses and how they are supposed to be installed or applied. Materials and installation or application failures and problems are identified and listed, then described in straightforward, understandable language supplemented by charts, graphs, photographs, and line drawings. Solutions ranging from proper cleaning and other maintenance and remedial repair to complete removal and replacement are recommended with cross-references to given problems.

Of further value are sections on where to get more information from such sources as manufacturers, standards setting bodies, government agencies, periodicals, and books. There are also national and regional trade and professional associations representing almost every building and finish material, most of which make available reliable, unbiased information on proper use and installation of their respective materials and products. Some associations even offer information on recognizing and solving problems for their products and materials. Names, addresses, and telephone numbers are included, along with each association's major publications. In addition, knowledgeable, independent consultants who specialize in resolving problems relating to certain materials are recognized. Where names were not available for publication, most associations can furnish names of qualified persons who can assist in resolving problems related to their products.

It is wished that one would never be faced with any problems with new buildings and even older ones. However, reality being what it is, this book, as do the others in the series, offers a guide so you can identify problems and find solutions. And it provides references for sources of more information when problems go beyond the scope of the book.

Jess McIlvain, AIA, CCS, CSI
Consulting Architect
Bethesda, Maryland

Preface

Architects working on projects where existing construction plays a part spend countless hours eliciting data from materials producers and installers relating to cleaning, repairing, and extending existing building materials and products, and for installing new materials and products over existing materials. The producers and installers know much of the needed information and generally give it up readily when asked, but they often do not include such information in their standard literature packages. As a result, there is a long-standing need for source documents that include the industry's recommendations for repairing, maintaining, and extending existing materials, and for installing new materials over existing materials. This book is one of a series called the *Building Renovation and Restoration Series* that was conceived to answer that need.

In the thirty-plus years I have worked as an architect, and especially since 1975, when I began my practice as a specifications consultant, I have often wondered why there is no single source of data to help architects, engineers, general contractors, and building owners deal with existing building materials. It is often necessary to consult several sources to resolve even apparently simple problems, partly because authoritative sources do not agree on many subjects. The time it takes to do all the necessary research is enormous.

I have done much of that kind of research myself over the years. This book includes the fruits of my earlier research, augmented by many additional hours of recent searching to make the book as broad as possible. In it, I have included as many of the industry's recommendations about working with existing acoustical materials, resilient flooring, and paint and transparent coatings as I could fit in. The data in this book, as is true for that in the other books in the series as well, come from the published recommendations of producers and their associations; applicable codes and standards; federal agency guides and requirements; contractors who actually do such work in the field; the experiences of other architects and their consultants; and from the author's own experiences. Of course, no single book could possibly contain all of the known data about those subjects or discuss every potential problem that could occur. Where data are too voluminous to include in the text, references are given to help the reader find additional information from knowledgeable sources. Some sources of data about historic preservation are also listed.

This book, as do the others in the series, explains in practical, understandable narrative, supported by line drawings and photographs, how to extend, clean, repair, refinish, restore, and protect the existing materials that are the subject of the book, and how to install the materials discussed in the book over existing materials.

All the books in the series are written for building owners, architects, federal and local government agencies, building contractors, university, professional, and public libraries, members of groups and associations interested in preservation, and everyone who is responsible for maintaining, cleaning, or repairing existing building construction materials. They are not how-to books meant to compete with publications such as *The Old House Journal* or the books and tapes generated by the producers of the television series, "This Old House."

I hope that, if this book doesn't directly solve your current problem, it will lead you to a source that will.

How to Use This Book

This book is divided into five chapters that discuss the subject areas suggested by their titles. Chapter 1 is a general introduction to the subject and offers suggestions as to how a building owner, architect, engineer, or general building contractor might approach solving the problems associated with dealing with existing building materials. It offers advice about seeking expert assistance when necessary, and suggests the type of people and organizations that might be able to help.

Each of the other four chapters includes:

- A statement of the nature and purpose of the chapter
- A discussion of materials commonly used to produce the support system (Chapter 2) or type of finish (Chapters 3 through 5) discussed in that chapter, and their usual uses
- A discussion of how and why those materials might fail. Chapters 2 through 5 include numbered lists of failure causes, such as "Selecting a system with too weak a structural classification for the conditions," that are grouped into failure source categories, such as "Structure Failure," "Structure Movement," "Improper System Design," "Bad Workmanship," and so on.

  In Chapters 3 through 5 there is also a list of failure types, such as "Stained or Discolored Acoustical Materials."

  Thus, a failure that is recognized in the field (stained acoustical panels, for example) can be traced to several possible causes, including, among many others in this example, "Poor Maintenance Procedures: Cause 1," which is "Failing to clean ceilings regularly." Where appropriate, there is also an explanation and discussion of the failure causes, so that a reader can see specific examples showing the effects of the error.

- A discussion of methods of extending and repairing materials that have failed and installing those types of materials over existing surfaces
- An indication of sources of additional information about the subjects in that chapter

The book has an Appendix that contains a list of sources of additional data. Sources include manufacturers, trade and professional associations, standards-setting bodies, government agencies, periodicals, book publishers, and others having knowledge of methods for restoring building materials. The list includes names, addresses, and telephone numbers. Sources from which data related to historic preservation may be obtained are identified with a boldface **HP.** Many of the publications and publishers of entries in the Bibliography are listed in the Appendix.

The items listed in the Bibliography are annotated to show the book chapters to which they apply. Entries which are related to historic preservation are identified with a boldface **HP.**

Building owners, engineers and architects, and general building contractors will, in most cases, each use this book in a somewhat different way. The following suggestions give some indication of what some of those differences might be.

Owners

It is probably safe to assume that a building owner is consulting this book because the owner's building has experienced or is now experiencing failure of one of the materials discussed here. If the problem is an EMERGENCY, the owner should turn immediately to Chapter 1 and read the parts there entitled "What to Do in an Emergency," and in "Help for Building Owners," the part called "Emergencies."

When the failure is temporarily under control, a more systematic approach is suggested. An owner may tend to want to turn directly to the chapter containing information about the finish that seems to have failed. An owner who has a good knowledge of such problems, and experience with them, may be able to approach the problem in that manner. Otherwise, it is better to first read and become familiar with the contents of Chapter 1, including those parts that do not at first seem to be applicable to the immediate problem. The cause of a finish failure is not always readily apparent, and jumping to an incorrect conclusion can be costly.

After reading Chapter 1, the owner should turn to the chapter covering the material that has failed. At some point, that chapter may refer to another chapter. Chapters 3, 4, and 5, for example, often refer to Chapter 2. When that happens, it is important to also read the cross-referenced material.

A word of caution, however. Unless experienced in dealing with such failures, an owner should not simply reach for the telephone to call for professional help, until after reading the chapter covering the failed material. It is always better to know as much as possible about a problem before asking for help.

Architects and Engineers

An architect's or engineer's approach will depend somewhat on the professional's relationship with the owner. For example, an architect who has been consulted regarding a finish failure may approach the problem differently depending on whether the architect was the existing building's architect of record, especially if the failure has occurred within the normal expected life of the failed finish. In such cases, there may be legal as well as technical considerations. This book, though, is limited to a discussion of technical problems.

An architect's or engineer's first impulse may be to rush to the site to determine the exact nature of the problem. For one who has extensive experience with finish failures, that approach may be reasonable. Someone with little such experience, however, should do some homework before submitting to queries by a client or potential client.

That homework might consist of reading the chapters of this book that deal with the apparent problem and consulting the sources of additional information recommended there. Then, if the problem is even slightly beyond the architect's or engineer's expertise, the next step is to read Chapter 1 and decide whether outside professional help is needed. An architect or engineer who has some related experience might delay that decision until after studying the problem in the field. One who knows little about the subject, however, will probably want someone knowledgeable to accompany them on the first site visit. Chapter 1 offers suggestions about how to go about making that decision.

An architect's or engineer's approach will be slightly different when they are commissioned to renovate an existing building. In that case, extensive examination of existing construction documents and field conditions is called for. When finish failures have contributed significantly to the reasons for the renovations, and the architect or engineer is not thoroughly versed in dealing with such conditions, it is reasonable to consider seeking professional assistance throughout the design process. In that event, the architect or engineer should read Chapter 1 first, then refer to other chapters as needed during the design and document production process. Even when there is a consultant on the team, the architect or engineer should have enough knowledge to understand what the consultant is advising and to know what to expect of the consultant.

Building Contractors

How a building contractor uses this book depends on which hat the contractor is wearing at the time and the contractor's expertise in dealing with existing building materials. For the contractor's own buildings, the suggestions given above for owners apply, except that the contractor will probably have more experience with such problems than many owners do.

When asked by a building owner to repair a failed finish, a contractor's approach might be similar to that described above for architects and engineers.

When repairs to a finish are part of a project for which a contractor is the general contractor of record, the problem is one of supervising the subcontractor who will actually repair the failed finish. Even a knowledgeable contractor will find it sometimes helpful to double-check the methods and materials that a subcontractor proposes against the recommendations of an authoritative source, such as those listed in this book. It is also sometimes useful to verify a misgiving that the contractor might have about specified materials or methods. In each of those cases, the contractor should read the chapter covering the subject at hand and check the other resources listed, whenever a question arises with which the contractor is not thoroughly familiar. Before selecting a subcontractor for repair work, a contractor might want to review Chapter 1.

Even a contractor who has extensive experience in repairing failed finishes will frequently encounter unusual conditions. Then the contractor should turn to the list of sources of additional information at the end of the appropriate chapter and the Appendix and Bibliography to discover who to ask for advice.

Disclaimer

The information in this book was derived from: data published by trade associations, standards-setting organizations, manufacturers, and government organizations, and statements made to the author by their representatives; interviews with consultants, architects, and building contractors; and related books and periodicals. The author and publisher have exercised their best judgment in selecting data to be presented, have reported the recommendations of the sources consulted in good faith, and have made every reasonable effort to make the data presented accurate and authoritative. However, neither the author nor the publisher warrant the accuracy or completeness of the data nor assume liability for its fitness for any particular purpose. Users bear the responsibility to apply their professional knowledge and experience to the use of data contained in this book, to consult the original sources of the data, to obtain additional information as needed, and to seek expert advice when appropriate.

Manufacturers and their products are occasionally mentioned in this book. Such mention is intended to indicate the availability of such products and manufacturers. No mention in this book implies any endorsement by the author or the publisher of the mentioned manufacturer or product, other products of the mentioned manufacturer, or any statement made by the mentioned manufacturer or associated in any way with the product, its accompanying literature, or in advertising copy.

Similarly, handbooks and other literature produced by various manufacturers and associations are mentioned. Such mention does not imply that the mentioned item is the only one of its kind available, or even the best available material. The author and publisher expect the reader to seek out other manufacturers and appropriate associations to ascertain whether they have similar literature, or will make similar data available.

Acknowledgments

A book of this kind requires the help of many people to make it valid and complete. I would like to acknowledge the manufacturers, producers' associations, standards-setting bodies, and other organizations and individuals whose product literature, recommendations, studies, reports, and advice helped make this book more complete and accurate than it otherwise would have been.

At the risk of offending the many others who helped, I would like to single out the following people who were particularly helpful:

Henry N. Carrera, CSI, Benjamin Moore and Company, Colonial Heights, Virginia

Margaret Ficklen, Association of the Wall and Ceiling Industries, International, Washington, D.C.

John W. Harn, Harn Construction Co., Laurel, Florida

Larry Horsman, National Concrete Masonry Association, Herndon, Virginia

Jerry D. Howell, Painting and Decorating Contractors of America, Peoria, Illinois

R. A. McDermott, CSI, Benjamin Moore and Company, Montvale, New Jersey

Bruce McIntosh, Portland Cement Association, Skokie, Illinois

Robert B. Molseed, AIA, FCSI, CCS, Professional Systems Division, AIA Service Corporation, Washington, D.C.

Helen Richards, Mid-City Financial Corporation, Bethesda, Maryland

Vincent R. Sandusky, Painting and Decorating Contractors of America, Falls Church, Virginia

Cathy Sedgewick, Association of the Wall and Ceiling Industries, International, Washington, D.C.

Sally Sims, Librarian, National Trust for Historic Preservation Library and University of Maryland Architectural Library, College Park, Maryland

Everett G. Spurling, Jr., FAIA, FCSI, Bethesda, Maryland

George H. Stewart, Sr., Stewart Brothers Photographers, Inc., Rockville, Maryland

CHAPTER
1

Introduction

A building's finishes are affected by many factors, including the construction of the building; the finish and support system materials themselves; the humidity and temperature during and after a finishes application; how the substrates receiving the finishes are prepared; the quality of the workmanship used to install the finishes and their support systems; and how well the finishes are protected after they have been installed. For interior finishes to remain in good condition, the building's shell and those elements, such as doors and windows, used to close openings in the shell must turn away wind and water and protect interior spaces from excessive temperature and humidity levels and fluctuations.

Almost every building will, at some time, experience a finish failure. Most of those failures will manifest themselves as cracks, discolorations, or disintegration in the finish material, or by the finish material separating from its substrate or other support system or changing its shape. When a finish failure occurs, there is both an immediate problem and a long-range problem.

The immediate problem, of course, is to stop the damage from getting worse. The long-range problem is to repair the damage in such a way that it will not recur.

This chapter includes a brief generic discussion of finishes and their failures, and outlines steps that a building owner can take to solve both the short- and long-range problems associated with failure. It also discusses the relationship of architects, engineers, general building contractors, specialty contractors, manufacturers, and damage (sometimes called forensic) consultants to the owner, and to each other, on projects involving finish failures, and outlines orderly ways in which those professionals can approach the problem-solving process.

This chapter includes an approach to determining the nature and extent of finish failures, and suggests the type of assistance that a building owner, architect, engineer, or contractor might seek to help solve the problem.

Chapters 2 through 5 contain detailed remedies and sources of additional data.

What Finishes Are and What They Do

The term *finish* in this book refers to applied materials used as the finished surface in and on buildings. This book includes acoustical ceilings, integrated ceilings, wall panels, and baffles; resilient flooring and associated products; paint and transparent finishes; and their support systems. Other finishes addressed in other books in this series include plaster; gypsum board; ceramic tile; nonstructural metals; and wood paneling and flooring. Materials which act as the final finish, but are often also part of the building shell or structure are addressed in other books in this series. They include stone, brick, and concrete unit masonry.

The term *failure* in this book refers to every type of failure from slight crazing of the surface to total disintegration of the material, and everything in between.

Failure Types and Conditions

A failed finish is often a symptom of an underlying problem, rather than a failure of the finish material itself. Recognizing that a finish failure has occurred often requires no special expertise (Fig. 1-1). Sagging ceilings, curling resilient flooring, and peeling paint, for example, are easy to see. Discovering the cause of the failure and determining the proper remedy,

Figure 1-1 No one could fail to recognize that there is a
paint problem here. (*Photo by Stewart Bros., courtesy of
Mid-City Financial Corporation.*)

however, often requires a detailed knowledge of finishes and their support
systems, and often of structural design as well.

When a finish fails, reviewing the appropriate chapter of this book, or
the other books in this series where the failed material is discussed, might
help an owner, architect, engineer, or building contractor identify the cause
of the failure and solve the problem. When, after reading the material
presented in this book series and examining the referenced additional data,
the reader does not feel competent to proceed without doing so, seeking
professional help is in order.

What to Do in an Emergency

In an emergency, it is often necessary to act first and analyze later. When action must be taken immediately to stop damage that is already occurring or is imminent, whatever is necessary should be done. Emergency action, for example, might consist of shoring a sagging ceiling using wood braces. If the failure is occurring because of a building or plumbing leak, it is, of course, necessary to stop the leak as soon as possible.

A word of caution is in order, however. Unless doing so is absolutely unavoidable, no irreversible remedial action should be taken. Small repairs that cannot easily be removed can become major cost items when permanent repairs are attempted. Hastily tearing holes in a ceiling to gain access to a leaking pipe could damage the ceiling to the extent that it must be completely removed later and a new ceiling installed.

Professional Help

If, after reading the chapters of this book or the other books in this series that address the identified problem area, the next step is still unclear, or it is not possible to be sure of the nature of the problem, consult with someone else who has experience in dealing with the kind of failure being experienced. Who that is will vary with the knowledge and experience of the person seeking the help.

Sometimes, obtaining more than one outside opinion is desirable. Certainly such is the case when a doubt exists about the opinion obtained. For those who have little experience in dealing with failed finishes, a second opinion is almost always desirable.

Help for Building Owners

Building owners can turn to several types of expert help when finish problems appear.

Emergencies. In emergencies, a building owner should seek help from the most available professional person or organization that can stop the damage from continuing. But, even in an emergency, it is better to seek help from a known organization than to simply thumb through the telephone book.

The first step is to call someone who can remedy the immediate problem—worry about protocol later. An exception to that rule should be made, however, if the building is still under a construction warranty. Then the owner should go to the person who is responsible for the warranty, which is usually a general contractor but may be a specialty contractor.

When the building is not under warranty, if the owner has a relationship with a building construction professional who is able to solve the immediate problem, that is the place to start. If the owner has relationships with several building professionals, the place to start is with the most appropriate. For example, if the failed finish was recently installed by a specialty contractor under direct contract with the owner—that is, there was no general contractor involved—the owner should call that specialty contractor.

Unless the person or firm is the only building professional that the owner knows, or the owner is unable to find someone to stop damage from continuing, calling an architect or engineer in an emergency is probably not appropriate. The best most of them will be able to do is to recommend an organization that might stop the damage from continuing. Going through them could waste valuable time.

In some areas of the country, a call to a local organization representing building industry professionals will net a list of acceptable organizations specializing in repairing the kind of damage that is occurring. Such local organizations might include a local chapter of the Associated General Contractors of America; an association of acoustical treatment, resilient flooring, or painting contractors, such as the Painting and Decorating Contractors of America; or the American Institute of Architects. Local chapters can be found by calling the affiliated national organization. Most of the applicable ones are mentioned in the appropriate chapters of this book. The names, addresses, and telephone numbers of the appropriate national organizations are listed in the Appendix.

It might be possible to find a local damage consultant who specializes in the kind of finish that has been damaged, but that person may only be able to recommend someone to repair the damage.

In each case, the methods that would be used under any other circumstance for contacting the various entities would apply in an emergency. There would not, however, be time to do the ordinary checking of references.

Architects and Engineers.   Even in nonemergency situations, building owners with easily identifiable finish problems often do not need to consult an architect or engineer. An exception occurs when repairing a finish failure is part of a general remodeling, renovation, or restoration project. Even when the needed repairs are solely evidenced by a finish failure, repairs requiring manipulation of structural systems, or which might cause harm to the public, may require a building permit. In many jurisdictions, building officials will not consider permit applications unless accompanied by drawings and specifications signed and sealed by an architect or engineer. Sometimes, the scope of the problem warrants hiring an architect or engineer to prepare drawings and specifications, even when authorities having jurisdiction do not require such documents. Occasionally, design considerations will make hiring an architect desirable whether or not such is required by law.

When the owner hires an architect or engineer, the same standards that one would usually follow when hiring such professionals apply. In addition to the usual professional qualifications applicable to any project, however, architects and engineers commissioned to perform professional services related to an existing building should have experience in the type of work needed. It is seldom a good idea to hire an architectural firm with experience solely in office building construction, no matter the scope, to renovate an existing hospital.

When a finish that has not reached its expected life span fails in a building that was designed for the owner by an architect, the owner should inform the architect of the failure and request assistance in solving the problem. In some cases, the architect will have a legal or contractual obligation to respond. Most architects will be interested enough to become actively involved, even when they are not required to do so by law or contract provision. Discussion of the architect's potential legal or contractual obligations in such cases is beyond the scope of this book. The architect may help determine the cause of the failure and suggest solutions. He or she may also help the owner determine which other professionals or manufacturers to contact and critique their recommendations. Reimbursement for the architect's services may or may not be appropriate, depending on the circumstances. Paint that peels slightly 15 years after it was applied, for example, is probably not the architect's responsibility under the original agreement between the owner and the architect.

General Building Contractors.   Finish failure problems that do not involve the building's structure or design are often best handled directly by a building contractor. Depending on the scope, the contractor may be a general contractor or a specialty contractor. When the problem is isolated and involves a single discipline, a single specialty contractor may be all that is needed. When multiple disciplines are involved, a general contractor may be needed to coordinate the activities of several specialty contractors. Owners with experience in administering contracts may be able to coordinate the work without a general contractor.

When selecting a general contractor, it is best to select one the owner knows. If there is no such contractor, the owner should seek advice in finding a competent firm. Such advice might come from satisfied building owners, architects, or engineers the owner knows, or the American Institute of Architects. As a last resort, an owner might ask for a recommendation from a contractors' organization. The owner should bear in mind, however, that asking for a recommendation from an organization that is supported by, and which represents, the firm being sought will, at best, get a list of firms in the area. That is not a good way to get a recommendation. This is in no way an indictment of contractors or their associations. The same advice applies to doctors, lawyers, and architects. You find the best ones by word of mouth.

If the project is large enough to warrant doing so, an owner might consider seeking competitive bids from a list of contractors. Few finish repair projects are that large or complicated, however. Negotiation is better, if reputable firms can be found with which to negotiate. If all of the parties are unknown, competitive bidding may be the only way to get a reasonable price.

When a finish fails in a building that is still under construction warranty, or the failed finish is still under a special warranty, the owner should contact the building contractor who is responsible for the warranty. The contractor's legal and contractual obligations under warranties may be complex and difficult to determine in specific cases. At any rate, they are beyond the scope of this book. Under ordinary circumstances, the owner should not have to initially contact the contractor's subcontractor or supplier for the failed finish, unless the general contractor cannot, or will not, do so. The owner should contact the architect and the owner's attorney for guidance about warranties and legal matters.

Even after the general and special warranties have expired, when a finish that has not reached its expected life span fails, the owner should probably contact the general contractor who built the building and request assistance in solving the problem. The owner might consider calling another contractor, if dissatisfied with the original contractor for some reason, but doing so may be costly. A new contractor will not be familiar with the building's peculiarities.

Specialty Contractors.   Specialty contractors and the associations that represent them, such as the Painting and Decorating Contractors of America, can sometimes be helpful in dealing with failed finishes. When a finish repair project is small and simple, it is often best for a building owner to deal directly with a specialty contractor. The methods suggested earlier for finding a general contractor may also be used to find a reputable specialty contractor. In addition, general contractors that the owner knows to be reputable are a source of recommendations for specialty contractors, so long as there is no symbiotic relationship between the general contractor and the specialty contractor. One of the best sources of the names of specialty contractors is the manufacturer of the product that is to be repaired.

Competitive bidding is an acceptable procedure when dealing with specialty contractors on small projects.

When a failed finish is under warranty, the specialty contractor who installed or applied the finish will probably be required to make the repairs, but the owner should not contact the specialty contractor directly when there was a general contractor involved in the original project. The general contractor is usually responsible for the warranty, and may choose to have repairs made by an entirely different subcontractor. There is normally no direct contractual relationship between the owner and the specialty contractor in that case.

When there is no general contractor, the owner should contact the specialty contractor, who is usually directly responsible for any warranty that may exist.

When the finish has failed because it was improperly installed or applied, and it is no longer under warranty, the owner may want to consider using a different specialty contractor to affect repairs or install a new finish.

Product Manufacturers.   Although some are more helpful than others, the manufacturer of the product that has failed, or an association representing the manufacturers of such products, are often knowledgeable sources of recommendations for dealing with finish failures. They may be the best-qualified agency to determine the cause of a failure. Some manufacturers have extensive training and testing procedures for their technical representatives, to ensure that they are qualified to deal with failures of the product. Training programs of four to six weeks are common. One paint manufacturer even conducts a psychological test to measure their representatives' social skill in dealing with disgruntled owners, architects, and contractors.

Some manufacturers employ national or regional representatives who are knowledgeable, trained, and equipped to determine the causes of failures and to recommend repair methods. One paint company, for example, equips its representatives with what they call a "complaint kit." It contains such diagnostic tools as a microscope, chemicals for determining a paint's composition, a micrometer to measure paint thickness, and knives and tweezers for taking samples.

In very simple cases, asking the manufacturer for advice may not be necessary. In more complex cases, however, contractors and architects will often bring in the failed finish manufacturer's representative to help determine the cause of, and solution to, a finish failure. When they do not, the owner should probably do so. When the failure is extensive, or its solution will be expensive or highly disruptive to the owner's activities, the owner may want to bring in a second manufacturer who makes a product similar to the failed one, to verify the advice given by the first manufacturer. Knowledgeable and reputable people often do not agree about the cause of failures and the methods needed to repair them and to prevent future failures. In some cases, the material itself may have failed, or the manufacturer's installation or application instructions may have been faulty. Then, the owner might be justified in not agreeing to throw good money after bad by further dealing with that manufacturer's products or advice. A new product or method may be needed to solve the problem.

Specialty Consultants. There are three sets of circumstances in which an owner may want to look for a consultant who specializes in the particular type of finish damage that has been encountered.

First, when a knowledgeable contractor or architect or a reputable and trustworthy manufacturer's representative is not available for consultation. Contractors and architects who have great knowledge about relatively recent materials and their failures may know little about the types of materials and systems used in older buildings. There may not be a manufacturer who has knowledge of a product that is no longer manufactured. A requirement for historic preservation may be sufficient cause to look for a specialty consultant.

Second, when the owner's contractor and architect and the manufacturer of the failed product cannot agree on the cause of a finish failure, the means appropriate to make repairs, or which party, if any, is responsible for the failure. For some failures, there may be no one to blame. No one can prevent finishes from being damaged by an earthquake, for example. For other failures, though, blame can often be placed. Battles about responsibility can be long and difficult to resolve. Too frequently, they end up in court. Sometimes, an outside "expert" can help resolve conflicts and prevent the parties from having to resort to litigation.

Third, when the people who are already involved cannot determine with certainty the cause of the damage or the proper method for making repairs.

Consultants who specialize in damage problems are sometimes called forensic consultants. They should be able to determine the nature of a problem and identify its true cause; find a solution to the problem; select the proper products to use in making repairs; write specifications and produce drawings related to the solution; and oversee the repairs.

Unfortunately, all such consultants are not created equal. Many who represent themselves as consultants are actually building product manufacturer's representatives trying to increase their sales or specialty contractors trying to enlarge their business. While most of them are reputable and some are competent to give advice, few are sufficiently knowledgeable to identify or advise an owner about solving underlying substrate or structure problems. Following the recommendations of an incompetent consultant can cause problems that will linger for years.

Selecting a consultant can be filled with uncertainty and the potential for harm. There are no licensing requirements, and no nationally recognized associations representing consultants who specialize in acoustical treatment, resilient flooring, or paint and transparent finish problems. Anyone who chooses to do so can hang up a shingle that says "Flooring Consultant," for example. As a result, a building owner who needs to hire a damage consultant must qualify that consultant with little help. One way to do this is to hire an architect or engineer and let them select and qualify the consultant, subject to the owner's approval, of course.

Asking a specialty contractor to recommend a consultant may not be a good idea. Although some of them hire damage consultants themselves, many specialty contractors feel that such consultants are, at best, a necessary evil. Their opinion probably stems from the tendency of some manufacturer's representatives to call themselves consultants and then oversell their abilities and knowledge.

Regardless of who makes the recommendation or does the hiring, the consultant should have a demonstrated expertise in dealing with the problems at hand. Obtaining references from the consultant's satisfied clients is an appropriate prequalification tool. A licensed architect or engineer who has extensive experience dealing with existing construction might be acceptable.

A consultant who is qualified to deal with finish failures in relatively new construction may know little about the kinds of materials that might be encountered in very old buildings. This is a particular problem when historic preservation is involved. A consultant for that type of work needs to demonstrate knowledge in dealing with very old materials and systems and the special requirements associated with historic preservation. It may not be possible to find a consultant who is an expert in dealing with old materials and also knowledgeable in general construction principles. In that case, it may be necessary to find two consultants with complimentary knowledge.

A damage consultant should have extensive knowledge of the material that has failed, in addition to an understanding of general construction principles and structure physiology. Many finish failures are a result of several problems. The consultant should know enough about buildings as a whole to be able to identify all of the underlying problems, and not just the obvious ones or those directly associated with the finish itself. Many finish problems are caused by structure movement, for example. The consultant should be able to determine whether the responsible movement is normal movement that has not been accounted for in the finish or its support system, or is a structure failure, which might require the repair of the structure itself. The consultant need not, however, be a structural engineer or know how to repair the structure. Usually, the ability to make the diagnosis is sufficient.

Finally, the consultant should have no financial stake in the outcome of the investigation. The owner needs to be sure that the consultant is an independent third party who is selling a professional service, and not the installation or repair of a product. Neither a product manufacturer's representative nor a contractor who wants to actually make the repairs meets this qualification.

Help for Architects and Engineers

Architects and engineers usually get involved in repairing finishes only when the failure occurred in a building they designed, or the repairs are extensive, have occurred on a prestigious building, or are part of a larger renovation, restoration, or remodeling project.

Architects who do not have extensive experience in dealing with finish problems or working with existing finishes should seek outside help from someone who has such experience. The nature of that help, and the person selected to consult, depends on the type and complexity of the problem.

Other Architects and Engineers.   One source of professional consultation for architects and engineers is other architects or engineers who have experience with the type of finish problem at hand. The qualifications needed are similar to those outlined earlier in this chapter for specialty damage consultants. The other architect or engineer need not be currently participating in an architectural or engineering practice. Specifications consultants and qualified architects and engineers employed by government, institutional, or private corporate organizations should not be overlooked.

Product Manufacturers and Industry Standards.   Manufacturers and their associations often provide sufficient information for a knowledgeable architect or engineer to deal with many finish failures. They are often the most knowledgeable source of data about failures of their products and related corrective measures.

An architect or engineer must, of course, compare product manufacturers' statements with those of other manufacturers and with industry standards, and exercise good judgment in deciding which claims to believe. The problem is the same as any other where architects and engineers consult producers and their associations for advice. The architect or engineer must study every claim carefully, especially if the claim seems extravagant, and double-check everything. Claims by the manufacturer of the failed product that identify causes other than product failure should be examined especially carefully. A second opinion may be helpful.

Also refer to "Product Manufacturers" under "Help for Building Owners" earlier in this chapter.

Specialty Consultants.   Architects sometimes employ independent specialty damage consultants, who may or may not be architects or engineers. The qualifications outlined earlier in this chapter for such consultants apply, no matter who is hiring the consultant.

Help for General Building Contractors

Whether a general building contractor needs to consult an outside expert depends on the complexity of the problem, the contractor's own experience, and whether the owner has engaged an architect or specialty consultant. Duplication of effort is unnecessary, unless the contractor intends to challenge the views expressed by the owner's consultants, or the content of the drawings and specifications.

A general building contractor acting alone on a project where the owner has not engaged an architect, engineer, or consultant must base the need for hiring consultants on such factors as the contractor's experience and expertise with the types of problems that will be encountered, the contractor's specialty subcontractor's experience and expertise in dealing with the types of problems involved, and the complexity of the problems.

Hiring a specialty consultant may complicate a general contractor's relationship with subcontractors. Such duplicity is seldom justified and is often a bad idea. An experienced and qualified specialty subcontractor is not likely to appreciate a damage consultant hired to tell the subcontractor how to do the job. The general contractor would be better off finding a qualified subcontractor to do the work and rely on that subcontractor's advice. If no such subcontractor is available, and the general contractor is not experienced with the problem at hand, hiring a consultant may be necessary, regardless of the feelings of the subcontractor. Muddling along to salve feelings is wholly incompetent and unprofessional. In those circumstances, the contractor should recommend that the owner employ a qualified architect or engineer to specify the repairs and let the owner and the design professional hire a damage consultant, if necessary.

In the event that a contractor should hire a specialty damage consultant, the recommendations earlier in this chapter for owners and architects apply.

It is often necessary for a general contractor who is working with an owner who has not hired an engineer or architect to consult with a failed product's manufacturer. The manufacturer's representative can often identify the causes of failure and recommend repair methods. When the contractor has engaged a specialty subcontractor, the manufacturer is often a source for a second opinion about such matters. The same precautions mentioned earlier for owners and architects dealing with a manufacturer apply when a contractor deals directly with a manufacturer.

Prework On-site Examination

On-site examinations before the work begins are important tools in helping to determine the type and extent of a finish failure and the damage to underlying construction it might portend. Who should be present during an on-site examination is dictated by the stage at which the examination will take place.

The Owner

The first examination should be by the owner or the owner's personnel to determine the general extent of the problem. This examination should help the owner decide what the next step should be and the type of consultant that the owner needs to contact, if any.

When the owner has selected a consultant, the owner and the consultant should visit the site and define the work to be done. The owner's consultant may be an architect or engineer, general contractor, specialty contractor, or specialty damage consultant. This second site examination should be attended by a representative of each expert that the owner has engaged to help with the problem. A specialty damage consultant, if engaged by another of the owner's consultants, should also be present. The general contractor's specialty subcontractors should also be present, if they have been selected.

During the second site visit, the parties should become familiar with conditions at the site and offer suggestions about how to solve the problem.

Architects and Engineers

An architect or engineer hired to oversee repair of a failed finish should, before visiting the site if possible, determine the products and systems used in the failed finish and the type of underlying and supporting construction. The architect should then visit the site with the owner to determine the extent of the work and to begin to decide how to solve the problem. Based on discussion with the owner about the nature of the problem, the architect should have decided whether to engage professional help or contact the failed product's manufacturer. If a consultant is to be used, that consultant should visit the site with the owner and the architect. If the manufacturer is to be consulted, its representative should accompany the owner and the architect on the site visit.

If, as a result of the architect's first site vist, the architect determines that a specialty damage consultant or manufacturer's representative, previously considered unnecessary, is needed, the architect should arrange for another site visit with that consultant or manufacturer's representative.

During the progress of the work, the architect, the architect's consultant, and the owner's representative should visit the site as often as is necessary to fully determine the nature and extent of the problem and to help arrive at a total solution. These site visits should extend the observation beyond the immediate problem to ascertain whether additional, unseen, damage might be present.

Building Contractors

Non-bid Projects. On non-bid projects, a building contractor may wear at least two hats.

The easiest situation to deal with is a negotiated bid based on professionally prepared construction documents. In that case, the contractor should conduct an extensive site examination to verify the conditions shown and the extent and type of work called for in the construction documents. Offering a proposal based on unverified construction documents is a bad business practice that can cost much more than proper investigation would have, if the documents are later found to be erroneous.

When the owner has not hired an architect or consultant to ascertain and document the type and extent of the work, the contractor must act as both designer and contractor. Then, the contractor should visit the site with the owner as soon as possible, and revisit as often as necessary, to determine the nature of the problem and the extent of the work to be done. The contractor may choose to hire a specialty consultant or specialty subcontractor, or both, or to consult with a representative of the manufacturer of the failed product. If so, those individuals should accompany the contractor on the site visits and participate in forming the contractor's recommendations. A carefully drawn proposal is an absolute must to be sure that the owner does not expect more than the contractor proposes to do.

Even when the owner hires an architect or other consultant, the contractor should visit the site with the owner and the owner's consultant as soon as possible. The purpose of the visit is to ascertain the extent and type of work to be done and to recommend repair methods. The contractor should also invite any specialty subcontractors to visit the site with the owner, the owner's consultants, the manufacturer's representative, and the contractor, if possible. The more input the contractor has in the design process, the better the result is likely to be.

Bid Projects. Even when the contractor is invited to bid on a project for which construction documents have been prepared, a prebid site visit is imperative. No contractor should bid on work related to existing construction without extensive examination of the existing building. Some construction contracts demand it. Some contracts even try to make failure to discover a problem the contractor's responsibility. Even if the courts throw that clause out, who can afford the time and costs of a lawsuit? A contractor should establish exactly what work is to be done before bidding. Insufficient data may be cause for not choosing to bid a project.

CHAPTER

2

Structure and Substrates

The finishes discussed in this book are supported by the building's structural framing system; solid substrates, such as concrete and unit masonry, which may be either structural or nonstructural; supported substrates such as plaster and gypsum board; and framing and furring systems erected specifically to support the finishes. Finish failures may be attributed to failure in any of those, and to problems with other building elements, such as roofing or wall construction.

This chapter includes a general discussion of structural systems, substrates, and other building elements, the failure of which may result in a failure of the finish. Framing and furring systems erected specifically to support finishes are addressed in the same chapter as the finish they support.

Refer to the book in this series entitled *Repairing and Extending Finishes, Part I* for additional information about the causes and means of correcting failures in the building's structural framing system, solid substrates, other building elements, and wood and metal framing and furring systems for plaster and gypsum board substrates; and for information about plaster and gypsum board substrates to which the finishes discussed in this book might be applied. Refer to other books in the series for additional data about finish failures due to weather barrier failure; masonry substrate failure; and wood framing and furring failure.

Structural system, solid substrate, supported substrate, and other building element problems often first appear as a failure in a finish. For the specific types of finishes discussed in each, Chapters 3 through 5 each contains a part called "Evidence of Failure," where a number of evidences of failure are listed. Each evidence of failure there is referenced back to other parts of that chapter ("Why Acoustical Ceilings Fail" in Chapter 3, for example) where the possible causes of the failure are listed. Among the failure causes listed in Chapters 3 through 5 are "Steel and Concrete Structure Failure," "Wood Structure Failure," "Structure Movement," "Solid Substrate Problems," "Supported Substrate Problems," and "Other Building Element Problems." Those six reasons for failure are addressed in this chapter and the numbers used for those reasons in "Evidence of Failure" in Chapters 3 through 5 correspond to the numbers listed for those failure causes in this chapter.

Structural Framing Systems

A building's structural system can have a major impact on whether the building's finishes remain in good condition or fail. The structural system can affect the finish in two ways: the structure can fail, or the structure can move in ways that are not anticipated in the finish design or installation.

Structures fail because they are improperly designed or because they experience conditions that exceed their design limitations. Design limitations are dictated by materials characteristics, legal requirements, and economic factors. It is not economically feasible to design every structure to handle every condition that might occur. Even a building designed to withstand an earthquake that measures 8 on the Richter scale may fail if the level reaches 8.5. It might not be possible to design a building that will be completely undamaged in even a small earthquake.

Structure failure may be small or large in scope, from slight damage to complete building collapse, but most structure failure is relatively small in magnitude. A single cause may generate a failure of any magnitude. An undersized footing, for example, may lead to building collapse or simply to more settlement than normal.

Since even minor structural failure may damage a building's finishes, it is necessary, before repairing failed finishes, to determine whether structural failure is responsible for the finish failing. Where structural failure is to blame, it is usually necessary to correct the structural failure before repairing the finish. Otherwise, the finish failure will usually recur. In severe cases, structural reconstruction, such as shoring up beams, adding columns, or replacing structural members may be necessary. When the structure damage is self-limiting and not dangerous to the building or to people, however, it is sometimes possible to modify the existing, or provide a new finish support system without making major corrections to the failed structure. An example is a self-limiting minor settlement caused by a small weak spot in the earth beneath a portion of a footing.

Repair of a failed structural system is beyond the scope of this book. Some such repairs are addressed in other books in this series, but this book has been written with the assumption that structural problems have been properly diagnosed and that necessary repairs have been made.

Steel and Concrete Structure Failure

Metal structural framing, such as steel beams, girders, trusses, bar joists, and decks are included in this category, as are cast-in-place and precast concrete structures.

Since the finishes addressed in this book are so seldom installed directly over such framing or furring, nonbearing metal wall and partition framing and furring, and metal stud-type framing designed to bear loads (cold-formed metal framing), are not included here. They are covered in some detail in the book in this series entitled *Repairing and Extending Finishes, Part I*.

Metal furring and suspension systems for acoustical treatment are discussed in Chapter 3.

Steel and concrete structure failure may occur if:

1. The structure is not properly designed to withstand all loads to be applied with no excess deflection, vibration, settlement (especially differential settlement), expansion, or contraction.

Wood Structure Failure

In addition to the general reasons for structure failure discussed above, wood structures have some additional potential problems.

The finishes discussed in this book are seldom applied directly on wood studs, columns, or posts. A plaster or gypsum board substrate is usually applied over wall framing to support acoustical treatment, for example. Therefore, construction of vertical wood structural framing is not discussed in detail in this book. Plaster and gypsum board, and the wood and metal framing and furring that supports them, are discussed in detail in the book in this series entitled *Repairing and Extending Finishes, Part I*. Wood and metal furring for acoustical treatment is discussed in Chapter 3 of this book. Wood and other subflooring for resilient flooring is discussed in Chapter 4.

Acoustical treatment is often suspended from wood floor and roof framing, however, including rafters, joists, beams, girders, trusses, and truss-joists. Less frequently, but often enough to consider here, acoustical treatment is secured directly to ceiling joists, roof rafters and trusses, or beams and girders. Then framing must provide an appropriate base by being level or in plane, stable, plumb, true, securely fastened in place, and at proper spacings.

Similarly, wood floor framing must be level, in proper plane, and correctly spaced to permit the proper application of subflooring to support resilient flooring.

The information in this book is not intended to give the reader enough information to build wood framing, but only to highlight those aspects of wood framing which, if not properly done, might cause applied finishes to fail. Refer to the sources mentioned in "Where to Get More Information" at the end of this chapter for detailed data about constructing wood framing.

Although there is no attempt here to discuss in detail the many possible wood building-framing systems, the general requirements related to stability, tolerances, and materials included here apply to the structural portions of the building when they support plaster or gypsum board substrates, acoustical treatment, or subflooring to receive resilient flooring. Wood framing should comply with the applicable building code and the standards and minimum requirements of generally recognized industry standards.

In general, framing design and member sizes, spacings, and locations should comply with recognized standards such as the National Forest Products Association's *Manual for Housing Framing*. Framing should be anchored, tied, and braced in such a way that it will develop the strength and rigidity necessary for its purpose. Members should not be spliced between supports.

Where pipes or other items interrupt the spacing of framing components, additional framing or furring must be provided to stabilize the framing and to properly support the applied acoustical treatment or subflooring.

Framing must be doubled or otherwise increased beneath unusual loads to provide a properly stable and level surface for the acoustical treatment or subflooring to be applied.

Bridging should be provided in floor and roof framing to ensure stability and to help prevent warp or twist in framing members.

Probably the single largest cause of problems with finishes associated with wood framing is the installation of improperly cured wood. Wood for framing should be seasoned lumber with 19 percent maximum moisture content at time of dressing. Lumber with a moisture content in excess of 19 percent can be expected to change in size by 1 percent for each 4 percent reduction in moisture content. As the wood changes in size, it will usually also warp and twist, especially when held in place at the ends, as is the case in framing members.

Softwood materials should comply with the U.S. Department of Commerce's *PS 20*, and the National Forest Products Association's *National Design Specifications for Wood Construction*. Each piece of lumber should be grade-stamped by an agency certified by the Board of Review, American Lumber Standards Committee.

Joists, rafters, planking, beams, stringers, and similar load-bearing members should be of at least the minimum grades required by the applicable building code and the engineering calculations made for the building.

In newer buildings, actual wood sizes will probably be in accord with the U.S. Department of Commerce's *PS 20*, and the lumber will probably be surfaced on all four sides. In older buildings, lumber may actually be 2 by 4 inches instead of 1-1/2 by 3-1/2 inches, as is common today. Or the actual dimension may be different from today's common actual dimension for the same nominal size. For many years, the normal actual thickness of a 2 by piece of lumber was 1-5/8 inches, and not 1-1/2 inches, as it is today, for example. Widths were also slightly larger. Older furring and framing lumber may not be surfaced on any sides, or may be surfaced only on the sides to which finishes are applied.

Wood roof trusses should have been designed for the manufacturer by a registered professional engineer to support the loads to be applied and in accordance with building code requirements. Roof trusses are factory fabricated, often from No. 2 Douglas fir, or equivalent. Connections are usually made with metal plate connectors or plywood gussets. Metal truss plates, whether pressed-in tooth or nail-in types, should comply with the Truss Plate Institute's "Design Specifications for Light Metal Plate Connected Wood Trusses." Truss plate connectors less than 1/8-inch thick should be zinc coated or noncorrosive metal and conform to the provisions of ASTM Standard A 525 for Commercial Coating Class.

Wood truss-type floor joists should also have been designed for the manufacturer by a registered engineer to support the loads to be applied and in accordance with building code requirements.

Wood should be pressure preservative treated where appropriate. Most framing will not be pressure preservative treated. Treated portions of framing usually include wood near or in contact with roofing or associated flashing; and wood sills, sleepers, blocking, furring, stripping, foundation plates, and other concealed members in contact with masonry or concrete.

Treated lumber should comply with the applicable requirements of the applicable standards of the American Wood Preservers Association (AWPA) and the American Wood Preservers Bureau (AWPB). Most recently treated framing lumber will have been treated in compliance with AWPA standards C2, using water-borne preservatives complying with AWPB standard LP-2. Some installations might be treated using other AWPA or AWPB recognized chemicals. Older applications may have been treated in accordance with other standards and using other chemicals, sometimes even creosote.

Lumber should be fire rated where required. Fire-retardant treated lumber and plywood should comply with AWPA standards for pressure impregnation with fire-retardant chemicals, and should have a flame spread rating of not more than 25 when tested in accordance with Underwriters Laboratories, Incorporated (UL) Test 723 or ASTM Standard E 84. It should show no increase in flame spread and no significant progressive combustion when the test is continued for 20 minutes longer than those standards require. The materials used should not have a deleterious effect on connectors or fasteners.

Treated items that are exposed to the exterior or to high humidities should be treated with materials that show no change in fire-hazard classification when subjected to the standard rain test stipulated in UL Test 790.

Fire-retardant treatment chemicals should not bleed through or adversely affect finishes applied to wood containing them.

Lumber with defects that might impair the quality of the finished surfaces, or that are too small to use in fabricating the framing or furring with the minimum number of joints possible or with the optimum joint arrangement, should not be used.

Framing or furring should be laid out carefully and set accurately to the correct levels and lines. Members should be plumb, true, and accurately cut and fitted. Openings should be framed and blocking provided for the related work of other trades.

Framing materials should be sorted so that defects have the least detrimental effect on the stability and appearance of the installation. Large or unsound knots should be avoided at connections. Materials at corners should be straight.

Framing should be securely attached to the substrates in the proper locations, and should be level, plumb, square, and in line. Anchoring and fastening should be done in accordance with applicable recognized standards. Nailing, for example, should be done in accordance with the "Recommended Nailing Schedule" in the National Forest Products Association's *Manual for House Framing*. Flat surfaces should not be warped, bowed, or out of plumb, level, or alignment with adjacent pieces by more than 1/8 inch in every 8 feet.

Nailing should be done using common wire nails. Fasteners should be of lengths that will not penetrate members where the opposite side will be exposed to view or will receive finish materials. Connections between members should be made tight. Fasteners should be installed without splitting of wood, using predrilling if necessary. Work should be braced to hold it in the proper position, nails and spikes driven home, and bolt nuts pulled up tight with heads and washers in contact with the work. Shims and wedges should be avoided.

Framing should be spiked and nailed using the largest practicable sizes of spikes and nails. The recommendations of the applicable recognized standards should be followed.

Framing that abuts vertical or overhead building structural elements should be isolated from structural movement sufficiently to prevent undesirable transfer of loads into the framing. Open spaces and resilient fillers are often used to fill spaces while preventing load transfer. Care must be taken to ensure that lateral support is maintained.

Both sides of control and expansion joints should be framed independently. The joints should not be bridged by framing or furring members.

Bolts, lag screws, and other anchors should be used to anchor framing in place. Generally, fasteners are placed near the top and bottom or ends of items and not more than 36 inches on center between. Shorter members, however, should be anchored at 30 inches on center.

Bolts should have nuts and washers.

Anchor bolts should be not less than 1/2 inch in diameter with the wall end bent 2 inches. They should extend not less than 8 inches into concrete and 15 inches into grouted masonry units. They should be placed at 48 inches on center with not less than 2 bolts in each member.

Expansion bolts should be not less than 1/2 inch in diameter and should be placed into expansion shields. The shields should be accurately recessed at least 2-1/2 inches into concrete.

Bolts in wood framing should be standard machine bolts with standard malleable iron washers or steel plate washers. Steel plate washer sizes should be about 2-1/2 inches square by 5/16-inch-thick for 1/2- and 5/8-inch diameter bolts, and 2-5/8 inches square by 5/16-inch-thick for 3/4-inch diameter bolts. Bolt holes in wood shall be drilled 1/16 inch larger than the bolt diameter.

Lag bolts should be square headed and of structural grade steel. Washers should be placed under the head of lag bolts bearing on wood.

Framing spacings will probably be dictated by structural considerations. If an acoustical treatment requires closer spacing than is appropriate for the structural considerations, cross furring should be used. Subflooring thickness should be determined in conjunction with the framing spacing so that the framing spacing and subflooring thickness are compatible.

Wood framing problems that can cause acoustical treatment, resilient flooring, or paint failures include the following:

1. Structure is improperly designed to withstand all loads to be applied

with no excess deflection, vibration, settlement (especially differential settlement), expansion, or contraction.

2. Misaligned, twisted, or protruding wood framing.
3. Lumber shrinkage. Even relatively dry wood will shrink.
4. Flexible or extremely hard wood framing.
5. Wet framing lumber.
6. Wood framing that is too dry when a directly attached acoustical treatment is installed.
7. Supports that are placed too far apart.
8. Failed framing. When the framing loses stability, the failure of an applied finish is inevitable.

Structure Movement

In addition to finish failures caused by structure failure, damage can occur because the structure moves, especially if the movement is larger than expected. If failure is to be prevented, finish design and installation or application must take structure movement into account. Unaccounted-for structure movement should be suspected when a finish fails, especially if the evidence of the failure is cracks or the finish separating from its substrates.

Undue structure movement may be a symptom of structural failure, as discussed in previous paragraphs. Some structure movement, however, is normal and unavoidable. Expected structural movement due to wind, thermal expansion and contraction, and deflection under loads is large in many modern buildings, which are purposely designed to have light, flexible structural frames that are less rigid than the structural systems used in most older buildings. Exterior column movement is a particular problem. Movement may be especially large in high-rise structures where both flexibility and wind loads are large. While these light modern designs are usually structurally safe, they are more likely to contribute to failures in finishes, unless the designer is aware of the problems they impose and takes steps to head off those problems. Finishes must be designed to accommodate the expected movement. Normal structure movement may be caused by one of the following:

1. Variable wind pressure, particularly on high-rise structures.
2. Structure settlement.
3. Thermal expansion and contraction.
4. Deflection of structural members and slabs.
5. Creep in concrete structures. Creep is a permanent change in the structure shape due to initial deflection in concrete elements.

6. Structure vibration, which is often transferred from operating equipment in the building. Vibration can loosen fasteners.
7. Lumber shrinkage. Even relatively dry lumber will shrink. Shrinking tends to cause lumber to warp or twist.
8. Lumber expansion due to the absorption of free water, condensation, or water vapor in high-humidity conditions.

Substrates

Substrates may be solid, such as concrete, unit masonry, and wood, which may be either structural or nonstructural; or supported, such as plaster, gypsum board, wood, and metal. Problems with either type may lead to finish failures.

Solid Substrate Problems

Solid substrates include those that are part of the building's structural system and those that are fillers, such as nonbearing walls and partitions.

Acoustical treatment is sometimes directly applied to concrete and masonry, but more commonly solid substrates to receive acoustical treatment are first furred and finished with a supported substrate, such as plaster or gypsum board. That is especially true for a directly applied acoustical treatment.

The most common solid substrate to which resilient flooring is applied is concrete. It may occasionally be directly applied to a solid wood substrate, but an underlayment is used over most other solid substrates. Refer to Chapter 4 for additional information.

Paint may be applied to almost any solid substrate, including concrete, unit masonry, wood, and stone.

Failures in solid substrates to which acoustical treatment, resilient flooring, or paint are applied either directly or over a supported substrate, such as plaster or gypsum board, will often damage the finish. The possibility of solid substrate failure must be investigated and ruled out before repairs to failed finishes are attempted. When a damaged solid substrate is responsible for a finish failure, it is often necessary to repair the solid substrate before repairing the finish. When the solid substrate damage is self-limiting and not dangerous to the building or to people, it is sometimes possible to repair an existing, or install a new, suspension, furring, or other support system to support the finish, without repairing the solid substrate.

The repair of solid substrates is beyond the scope of this book. This book has been written with the assumption that damaged solid substrates have been discovered and that necessary repairs have been made. The reader should recognize that the construction and finishing methods used in constructing solid substrates may drastically affect applied finishes and contribute to any failure that occurs.

Solid substrate problems that can lead to finish failure include the following:

1. The solid substrate exudes materials that affect the finish or cause it to delaminate from the solid substrate. Some substances extruded by solid substrates that can cause harm to applied finish materials are not foreign to the substrate material. It is perfectly natural for a concrete wall or slab to evaporate water for an extended period, for example. A slab should be permitted to dry for at least 60 days before resilient flooring is installed, and some flooring manufacturers recommend a longer drying time. Concrete and masonry must cure completely before most paint can be successfully applied. Lightweight concrete will take longer to dry than will normal-weight concrete. Additives can increase drying time, as can high humidity conditions. Slabs placed on the ground take longer to dry than do supported slabs with ventilated space below them. Using a low water-cement ratio concrete, on the other hand, will reduce the amount of evaporable water and decrease the drying time. Applying a finish too early can lead to its failure. Failing to wait long enough before applying a finish is actually a bad workmanship problem, however, rather than substrate failure.

   Some foreign substances that will cause finish failure are the result of bad design or workmanship related to the substrate and have nothing to do with the finish installer's workmanship. Capillary moisture migrating from the underlying soil through a concrete slab-on-grade is usually caused by bad design or workmanship related to the underlying structure, for example. Such migrating moisture can collect beneath the finish flooring and cause it to loosen or pop up. Dissolved alkalies (mineral salts) from the concrete can be carried upward with the migrating water and collect on the surface of the substrate and chemically attack the flooring's adhesive, causing the flooring to delaminate from the concrete. A more watertight concrete can be achieved by moist-curing slabs for the first five days of the curing cycle. Some errors that cause or permit water migration to occur are discussed under "Other Building Element Problems," later in this chapter.

   Efflorescence on masonry or concrete walls is another example of a usually preventable substrate problem. The most common form of efflorescence is a white powdery deposit of alkali salts and carbonates, but other salts sometimes appear as well. Efflorescence on brick masonry may be green vanadium salts or brown, tan, or gray manganese dioxide salts, for example.

The Brick Institute of America, National Concrete Masonry Association, and Portland Cement Association publications listed in "Where to Get More Information" at the end of this chapter contain those organizations' recommendations related to maintaining and cleaning the applicable solid substrates and preventing or curing the types of problems discussed here.

2. The solid substrate material cracks or breaks up, joints crack, or surfaces spall due to bad materials, incorrect material selection for the location and application, or bad workmanship.

3. The solid substrate moves excessively due to improper design or installation. Excess movements that can cause problems include deflection, vibration, settlement (especially differential settlement), expansion, and contraction.

4. Unaccounted-for normal movement. As is true for structural systems, some movement in solid substrates is normal and unavoidable. That movement must be accounted for in the finish. Normal movement includes that caused by settlement, thermal expansion and contraction, creep, and vibration.

Supported Substrate Problems

The most common supported substrates to which acoustical treatment and paint are directly applied are plaster and gypsum board. In new buildings, acoustical treatment is rarely installed directly (without furring) over other supported substrates. In renovation and remodeling projects, however, acoustical treatment may be installed over almost any existing material, including such supported substrates as wood paneling and ceramic tile.

Resilient flooring may be applied over wood finish flooring, wood subflooring, ceramic tile, and other floor finishes. Refer to Chapter 4 for additional information about those substrates.

Paint may be applied to any supported substrate, including plaster, gypsum board, wood, plywood, metal, cement composition board, tile, and even glass.

It is not the purpose of this book to discuss supported substrate construction methods. The reader should recognize, however, that the construction and finishing methods used on supported substrates may drastically affect applied finishes and contribute to any failure that occurs.

Failures of supported substrates to which finishes are applied either directly or with furring will often damage the finish (Figs. 2-1 and 2-2). The possibility of supported substrate failure must be investigated and ruled out before repairs to finishes are attempted. When a damaged supported substrate

Figure 2-1 A crack in the supporting gypsum board caused the paint failure on the jamb at this apartment window. (*Photo by author.*)

is responsible for a finish failure, it is often necessary to repair the substrate before repairing the finish. When the substrate damage is self-limiting and not dangerous to the building or to people, it is sometimes possible to repair an existing, or install a new, suspension, furring, or other support system to support the finish without repairing the substrate.

Repair of supported substrates is beyond the scope of this book, which has been written with the assumption that damaged substrates that are discovered have been repaired when necessary. Plaster, gypsum board, and ceramic tile, along with framing and furring for them, are discussed in the book in this *Building Renovation and Restoration Series* that is entitled *Repairing and Extending Finishes, Part I.* Many of the following numbered

Figure 2-2 This paint failure is due to rotting away of the underlying wood window sill. (*Photo by author.*)

failure causes, and the reasons they occur, are discussed in detail in that book. In that book, the numbered failure causes below are called failure types.

Supported substrate problems that can lead to acoustical treatment or paint failure include the following:

1. Loose plaster.
2. Soft or crumbling plaster.
3. Cracking plaster.
4. Plaster blisters.
5. Stains on plaster substrates.
6. Efflorescence on plaster substrates.
7. Grease or silt embedded in plaster.
8. Wavy gypsum board surfaces.
9. Loose or sagging gypsum ceiling boards.
10. Soft or disintegrating gypsum boards.
11. Gypsum board cracks.
12. Gypsum board joint cracks.
13. Stains on gypsum board.
14. Gypsum board joint beading or bridging.

15. Blisters on gypsum board surfaces or joints.
16. Gypsum board joint compound debonding.
17. Failing to properly prepare plaster or gypsum board substrates. The Gypsum Association recommends that gypsum board surfaces to receive high-gloss paint be coated with a skim coat of joint treatment material or that a veneer plaster be used instead. Excessive sanding of plaster or gypsum board paper surfacing or joint treatment can result in differential suction between sanded and unsanded areas, resulting in a lack of continuity in an applied paint's color or sheen.

Other Building Elements

Other building elements that are poorly designed, or that fail, can cause finishes to fail. Those other building elements include site grading adjacent to the building; capillary water barriers, moisture retarders, waterproofing, insulation, and drainage systems beneath concrete slabs; openings, such as windows, doors, and louvers; caulking and sealants; and mechanical and electrical systems.

Other Building Element Problems

The types of failures in other building elements that can cause finish failure include the following:

1. Other building elements, whether poorly designed or simply failed, that permit water to enter the finish, its substrates, or both, or to wash down the surface of the finish. Possible sources of water intrusion or washing include roof and plumbing leaks (Figs. 2-3 and 2-4); failed sealants; leaks through doors, windows, louvers, and other openings; and leaks through, or missing, rain gutters and downspouts.

    Water occurring below a concrete slab-on-ground can migrate up through the slab by capillary action, unless some action is taken to prevent the occurrence. Water may be present because the exterior finish grade does not properly drain water away from the building; because the water table is high and a proper drainage system has not been installed to pipe the water away from the building; because capillary action has not been prevented by a capillary barrier, consisting of at least a 4-inch-thick granular fill or a sheet drainage medium and a layer of vapor retarder or waterproofing (which is used depends on the amount of water present and its hydrostatic pressure).

    Water entering exterior finish materials can contribute to damage

Figure 2-3 A plumbing leak damaged this acoustical panel. (*Photo by author.*)

from mildew or other fungal growth; moss, ivy, or other plant growth; and paint blistering, peeling, flaking, cracking, or scaling.

2. Designs that permit condensation to form and enter the finish, its furring, suspension system, or substrates. Condensation can result from selecting the wrong materials or installation methods for insulation and vapor retarders, or improperly locating those elements. Failing to provide proper ventilation in attic spaces can also lead to condensation affecting finishes. Condensation on the surface of concrete slabs-on-ground can sometimes be prevented by imposing a layer of insulation beneath the slab to prevent the concrete from reaching the dew-point temperature.

3. Placing unsupported insulation directly on acoustical ceiling materials or suspension systems. Some ceiling suspension systems will support insulation loads, but the condition should be checked with the finish manufacturer when a failure occurs that might be related to applied loads.

4. Supporting lighting fixtures, ducts, pipes, conduit, or other mechanical or electrical equipment or devices on acoustical ceiling materials or suspension systems that are not designed to carry the load. This

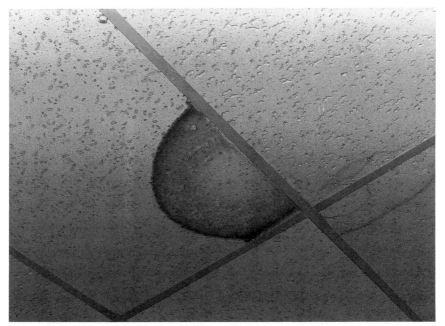

Figure 2-4 A roof leak stained this acoustical ceiling. (*Photo by author.*)

is particularly a problem when the suspension system is rated for light duty, but may apply to other systems as well.

5. Temperature differential between interior and exterior materials. When the framing or furring, or the fasteners attaching a supported finish to the framing or furring, are colder than the face of the finish, the finish directly over the cold elements will be colder than adjacent surfaces. Airborne dust particles will be attracted to the colder areas and stick there. The result will be dark spots, areas, or lines directly beneath the cold framing, furring, or fasteners. The dark dust accumulations are called shadowing.

Where to Get More Information

The National Forest Products Association's *Manual for House Framing* contains a comprehensive nailing schedule and other significant data about wood framing. It is a useful tool for anyone who must deal with wood construction of any type.

The Forest Products Laboratory's *Handbook No. 72—Wood Handbook* contains a detailed discussion of wood shrinkage.

Some of the AIA Service Corporation's *Masterspec* Basic sections contain excellent descriptions of the materials and installations that are addressed in this chapter. Unfortunately, those sections contain little that will help with troubleshooting failures in the materials and systems addressed in this chapter. Sections that have applicable data are marked with [2] in the Bibliography.

Every designer should have the full complement of applicable ASTM Standards available for reference, of course. Some of those applicable to the materials and systems discussed in this chapter are marked with a [2] in the Bibliography. They may also be mentioned in this and other chapters.

The following Brick Institute of America's *Technical Notes on Brick Construction* contain BIA's recommendations related to subjects discussed in this chapter:

The September/October 1977 edition of technical note "20, Cleaning Brick Masonry" contains BIA's recommendations for cleaning new and existing brick masonry, including, but not limited to, removal of efflorescence and other stains.

The December 1969 edition of technical note "23, Efflorescence, Causes" and the January 1970 edition of technical note "23A, Efflorescence, Prevention and Control" contain information about the causes of different types of efflorescence and BIA's recommendations for preventing and controlling it.

The following Portland Cement Association publications contain PCA's recommendations related to subjects discussed in this chapter:

The 1965 "Bulletin D89: Moisture Migration—Concrete Slab-On-Ground Construction."

The 1977 publication "Painting Concrete, IS134T."

The 1982 publication "Removing Stains and Cleaning Concrete Surfaces."

The 1983 publication "Concrete Floors on Ground."

The 1986 publication "Effects of Substances on Concrete and Guide to Protective Treatments."

The National Concrete Masonry Association's 1972 edition of "NCMA-TEK 44—Maintenance of Concrete Masonry Walls" contains NCMA's recommendations on subjects discussed in this chapter.

The Gypsum Association's 1985 publication "Using Gypsum Board for Walls and Ceilings" (GA-201-85) includes GA's recommendations for finishing gypsum board that is to be decorated with paint.

Also refer to other items marked [2] in the Bibliography.

CHAPTER

3

Acoustical Treatment

Acoustical treatment discussed in this chapter includes acoustical ceilings; integrated ceilings; and acoustical wall treatment and baffles. Furring and suspension systems associated with acoustical treatment are also discussed in this chapter. Other construction associated with acoustical treatment, such as the building structural system, solid and supported substrates, and other building elements, are discussed in Chapter 2.

Before proceeding, it will be helpful to define some of the terms we will use in this chapter. When this chapter uses the following terms, the listed definitions apply.

Acoustical Baffles: An acoustical panel that is suspended from an edge or which floats cloud-fashion in a space is called an acoustical baffle.

Acoustical Ceiling: The entire ceiling system, including the acoustical material and associated insulation, furring, and suspension systems, is called an acoustical ceiling. This term does not include auxiliary thermal or acoustical insulation in addition to that normally associated with the acoustical ceiling being discussed. Thus, a layer of acoustical insulation placed above a ceiling at a location over, or adjacent to, a partition to achieve increased acoustical performance is not here included as part of the acoustical ceiling. An acoustical blanket or pad which is necessary to achieve the rated acoustical performance of a metal ceiling is, however, a part of the acoustical ceiling. Metal ceilings, as defined below, may also be called acoustical ceilings in this chapter.

Acoustical Material: The general term acoustical material includes both composition and metal acoustical materials, as defined in this chapter.

Acoustical Wall Panels: Acoustical panels that mount on the surface of a wall or partition are called acoustical wall panels, regardless of their mounting method or size. Acoustical wall panels may completely cover a wall surface or be mounted as plaques with space between them.

Composition: When referring to acoustical materials, the term composition includes cellulose, mineral fiber, glass fiber, asbestos cement, ceramic-faced, plastic, gypsum, and wood fiber acoustical materials. Plastic- and metal-faced acoustical ceiling materials in which the plastic or metal is adhered or bonded to a composition material, as defined in this paragraph, are also called composition materials.

Integrated Ceiling: A ceiling system consisting of acoustical material, a suspension system, air distribution outlets, and lighting fixtures.

Metal Acoustical Ceiling Material: Linear metal strips, and metal panels and pans, are all called metal acoustical materials, even though some of them may be of questionable acoustic value. The term metal acoustical material means the ceiling material only. It does not include the associated suspension system or acoustical insulation blankets or pads.

Metal Ceiling: The entire ceiling construction, including metal acoustical materials, acoustical pads or blankets, and the suspension system, is called a metal ceiling.

Acoustical Ceilings

Acoustical ceilings include composition tiles and panels; metal acoustical materials; and associated furring and suspension systems. Integrated ceilings are discussed later in this chapter.

Acoustical Materials

The characteristics of acoustical materials are generally identified according to Federal Specification SS-S-118, which classifies them according to types, pattern, class (flame spread index), grade (Noise Reduction Coefficient [NRC] or Speech Privacy Noise Isolation Class [NIC']), and light reflectance coefficient. Types include the following:

Type I   Cellulose composition with standard washable painted finish

Type II   Cellulose composition with plastic membrane-faced overlay

Type III   Mineral composition with standard washable painted finish, which includes:
            Form 1 Nodulated, cast, or molded
            Form 2 Water felted
            Form 3 Dry felted

Type IV   Mineral composition with plastic membrane-faced overlay, which includes:
            Form 1 Nodulated, cast, or molded
            Form 2 Water felted
            Form 3 Dry felted

Type V   Steel facing with mineral composition absorbent backing

Type VI   Stainless steel facing with mineral composition absorbent backing

Type VII   Aluminum facing with mineral composition absorbent backing

Type VIII   Cellulose composition with scrubbable pigmented or clear finish

Type IX   Mineral composition with scrubbable pigmented or clear finish, which includes:
            Form 1 Nodulated, cast, or molded
            Form 2 Water felted
            Form 3 Dry felted

Type X   Mineral composition with plastic/aluminum membrane

Type XI   Other (this category is left for additional types)

Until recent years, SS-S-118 Type VIII was "Asbestos cement board facings with mineral composition absorbent backing." Because of the realization of the dangers associated with asbestos, that type is no longer acceptable. It may, however, occur in an existing building.

Patterns are divided by Federal Specification SS-S-118 into the following:

a. Regular large hole perforated
b. Randomly large hole perforated

c. Finely perforated
d. Fissured
e. Textured light to medium
f. Textured heavy
g. Smooth
h. Printed
i. Embossed
j. Embossed-in-register
k. Other (this category is reserved for special finishes)

Federal Specification SS-S-118 further divides acoustical materials into three classes for flame spread: Class A material ranges from zero to 25 flame spread index; Class B from 26 to 75; and Class C from 76 to 200.

Grade, which refers to acoustical performance, is rated either by Noise Reduction Coefficient (NRC) or Speech Privacy Noise Isolation Class (NIC').

Light reflectance is rated as one of four coefficients: LR 1 includes light reflectances of 75 percent or more; LR 2 ranges from 70 to 74 percent; LR 3 from 65 to 69 percent; and LR 4 from 60 to 64 percent.

Materials and Finishes.   Existing acoustical ceiling materials may be any of the types mentioned above, in any of the patterns or grades mentioned, and with any of the light reflectance coefficients mentioned.

Other materials may also be encountered, including ceramic-faced units that are used in wet and high-impact areas; units with other abuse-resistant or sanitary plastic faces; gypsum; and wood fiber composition. Older acoustical units may contain asbestos. Suspended ceilings may also be of wood, plastic, or sheet metal, but most of those types are more properly classified as decorative, rather than acoustical, ceiling materials.

Most composition acoustical units are painted on the exposed face. Most such painted finishes are considered washable, although gentleness is advised. Some manufacturers list some products as having a "scrubbable" finish, but there is no universally accepted definition of the term.

Metal ceiling panels are available in aluminum, steel, and stainless steel. A variety of finishes are available including fluoropolymer, baked enamel, natural, anodized aluminum, acrylic lacquer, and even wood veneer.

Acoustical Performance.   Acoustical products are classified either by Sound Transmission Class (STC) and Noise Reduction Coefficient (NRC), or by Speech Privacy Noise Isolation Class (NIC'). STC and NRC ratings are usually used in closed-room situations. NIC' is used for large spaces that contain several activities, such as occurs in an open-plan office space. NIC'-rated products have been tested for their ability to both control sound reflections and uniformly permit white-noise transmission from plenum speakers into a space. White-noise is used in open space designs to mask background noises and help provide speech privacy.

The sound absorption quality of an acoustical material is indicated by its NRC, which shows a material's ability to reduce sound levels within a space. An acoustical material's noise isolation control ability is indicated by the STC, which shows a material's ability to prevent sound from passing through it into another space. Usually there is an inverse relationship between the two properties. Materials that reduce sound reverberations are usually light and porous. To prevent sound transmission, materials must be heavy and airtight. The STC is measured by test procedure AMA 1-II "Ceiling Sound Transmission Test by the Two Room Method," which gets its name from the no-longer-existent Acoustical Materials Association. The standard is available from the Ceilings and Interior Systems Contractors Association (CISCA).

Where sound control is an essential element in an existing building, the advice of an acoustical specialist should be sought before changes or repairs are made.

Composition Acoustical Tile versus Panels. Composition acoustical ceiling materials may be either tile or panels. In the past, much tile was used. In recent years, the advent of better acoustical panel materials, more attractive and functional suspension systems, an increased need for access to the space above ceilings, changes in construction methods, their lower cost, and other factors has led to a much greater use of panels. Some manufacturers have stopped making tile products and concealed grid suspension systems. An advantage of panels over tile is that panels are loosely laid into a grid, which makes them easily removable and makes the space above them readily accessible. A major disadvantage is that they are loosely laid into a grid which makes them subject to dislocation by wind, pressure applied during cleaning, and air passing through a plenum space above them. Tile and panel edges are square, beveled, or have a special edge, such as the reveal edge shown in Figure 3-1.

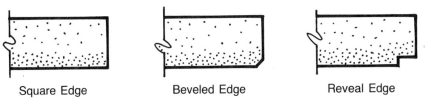

  Square Edge   Beveled Edge   Reveal Edge
Figure 3-1 Acoustical tile and panel edge conditions.

Tile. Any of the types, patterns, and other characteristics listed in previous paragraphs may be found. Older tile will probably be 12, 16, or 20 inches square. Newer tile is more likely to be 12 or 24 inches square or 12 by 24 inches.

Tile joints are flanged, rabbeted, kerfed (splined), or tongue and groove. Figure 3-2 shows some types, but there are many others in use today, and still others have been used in the past. Sometimes, a tile's edges are not all cut the same. The tongue and groove joints on two sides of a tile may be quite different from those on the other two sides, so that the tile will nest properly.

Panels. Any of the types, patterns, and other characteristics listed in previous paragraphs may be found in panels. Most panels are either 24 inches square or 24 by 48 inches. Some are scored so that they appear from the floor to be smaller units. Scoring may produce squares or linear strips. Scoring may be parallel with the tile sides or diagonal to them.

Panel joints are usually square or reveal type (see Fig. 3-2). In systems where cross-suspension members are concealed (semiconcealed systems), the ends of the panels are usually kerfed and rabbeted to receive the concealed suspension member.

Metal Acoustical Ceiling Materials.   In addition to metal-faced composition products, there are many metal acoustical ceiling material styles. They

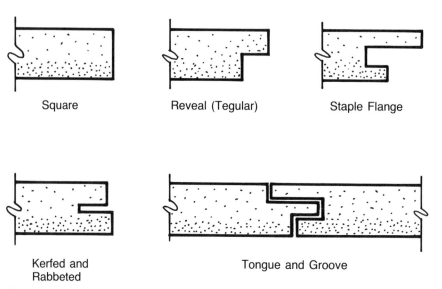

Square   Reveal (Tegular)   Staple Flange

Kerfed and   Tongue and Groove
Rabbeted

Figure 3-2 Typical acoustical tile joint conditions.

include, but are certainly not limited to, linear shapes (strips), flat panels, shaped (corrugated and other shapes) panels, and flat or textured pans. Surfaces may be continuous or perforated, and either flat or embossed. Metal grids and louvers are also available, but whether they are acoustical in nature is debatable. Metal ceiling members may be either the lay-in type or snap-in type. Snap-in types require special support systems. Some lay-in types also require special support systems, while others do not. Some panels come with glass fiber or mineral wool sound-absorbing pads, and others do not.

Support Systems

Composition acoustical materials may be directly attached to solid or supported substrates, or supported on a suspension system. Metal ceilings usually include a suspension system. Generally, the term "suspension system" means some sort of metal grid that is hung by wires or bars from the structure, but the term also includes systems of wood or metal furring to which acoustical materials are screwed, stapled, or nailed. Composition materials so attached are usually tiles. Sheet metal decorative ceilings are often held in place using fasteners, but metal acoustical ceiling materials are seldom so installed.

Wood Furring.   Except in single family residences and other small buildings, little acoustical tile is installed directly on wood furring today (Fig. 3-3),

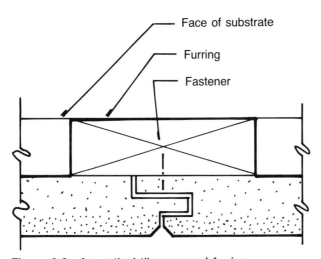

Figure 3-3 Acoustical tile on wood furring.

but in the past such installations were common. The normal installation when wood furring is used today is to install a layer of gypsum board, called backing board, over the furring and fasten the acoustical tile to the gypsum board. Wood framing and furring that supports plaster and gypsum board are discussed in detail in the book in this series entitled *Repairing and Extending Finishes, Part 1.*

When wood furring is used, it should comply with the building code and the standards and minimum requirements of generally recognized industry standards.

Wood.   Probably the single largest cause of problems with acoustical ceilings applied directly to wood furring is the installation of improperly cured wood. Wood for furring should be seasoned lumber with 19 percent maximum moisture content at the time of dressing. Lumber with a moisture content in excess of 19 percent can be expected to change in size by 1 percent for each 4 percent reduction in moisture content. As the wood changes in size, it will usually also warp and twist, especially when held in place at the ends, as is the case in framing and furring members.

Softwood used for furring should comply with the U.S. Department of Commerce's *PS 20,* and the National Forest Products Association's *National Design Specifications for Wood Construction.* Each piece of lumber should be grade-stamped by an agency certified by the Board of Review, American Lumber Standards Committee.

Lumber for furring may be any of a number of available species, including, but not limited to, Douglas fir, Douglas-fir-larch, hem-fir, southern pine, spruce-pine-fir, and redwood.

In newer buildings, actual wood sizes will probably be in accord with the U.S. Department of Commerce's *PS 20,* and the lumber will probably be surfaced on all four sides. In older buildings, though, lumber may have a different actual dimension than is normal today for the same nominal size. A 2 by 4, for example, may actually be 2 inches by 4 inches instead of 1-1/2 by 3-1/2 inches, which is the standard today. Or, it may be 1-5/8 by 3-5/8 inches, which was the standard for many years. Older furring and framing lumber may not be surfaced on any sides, or may be surfaced only on the sides to which finishes are applied.

Most furring that directly supports acoustical ceilings will not be pressure preservative treated. Furring that may be treated includes that near or in contact with roofing or associated flashing. Sometimes furring in contact with masonry or concrete will also be pressure preservative treated.

Treated lumber should comply with the applicable requirements of the applicable standards of the American Wood Preservers Association (AWPA) and the American Wood Preservers Bureau (AWPB). Each treated item should be marked with an AWPB "Quality Mark," but the markings may not be visible after installation.

Most recently-treated furring will have been treated in compliance with AWPA standard C2, using water-borne preservatives complying with AWPB standard LP-2. Some installations might be treated using other AWPA or AWPB recognized chemicals. Older applications may have been treated in accordance with other standards and using other chemicals, sometimes even creosote.

Most furring that directly supports acoustical ceilings will not be fire-retardant treated. Furring in walls and floors where fire-rated construction is required by the building code may be treated, however.

Fire-retardant treated lumber should comply with AWPA standards for pressure impregnation with fire-retardant chemicals, and should have a flame spread rating of not more than 25, when tested in accordance with Underwriters Laboratories, Incorporated (UL) Test 723 or ASTM Standard E 84. It should show no increase in flame spread and no significant progressive combustion when the test is continued for 20 minutes longer than those standards require. The materials used should not have a deleterious effect on connectors or fasteners.

Treated items that are exposed to the exterior or to high humidities should be treated with materials that show no change in fire-hazard classification when subjected to the standard rain test stipulated in UL Test 790.

Fire-retardant treatment chemicals should not bleed through or adversely affect the acoustical ceiling materials used.

Each piece of fire-retardant treated lumber should have a UL label, but the labels may not be visible after application.

Miscellaneous Materials. The following miscellaneous materials are necessary for wood furring installation:

- Rough hardware, metal fasteners, supports, and anchors.
- Nails, spikes, screws, bolts, clips, anchors, and similar items of sizes and types to rigidly secure furring members in place. These items should be hot-dip galvanized or plated when in contact with concrete, masonry, or pressure-treated wood or plywood, and where subject to high moisture conditions or exposed.
- Suitable rough and finish hardware as necessary.
- Bolts, toggle bolts, sheet metal screws, and other suitable approved anchors and fasteners. These should be located not more than 3 feet on center to firmly secure wood furring in place. Nuts and washers should be included with each bolt. Anchors and fasteners set in con-

crete or masonry should be hot-dip galvanized, and should be types designed to be embedded in concrete or masonry as applicable and to form a permanent anchorage.

Metal Furring.   Today, very little acoustical material is installed directly on metal furring bars (Fig. 3-4), but in the past some such installations did occur. Today, when composition acoustical tile is installed over metal furring bars, the method usually used is that described in ASTM Standard C 635 as "Furring Bar Suspension System," in which a layer of gypsum board, called backing board, is applied to suspended nailing or furring bars. The acoustical tile is then fastened to the gypsum board. Metal framing and furring that supports plaster and gypsum board are discussed in detail in the book in this series entitled *Repairing and Extending Finishes, Part I.* Metal acoustical materials are seldom installed over metal furring bars today.

Metal acoustical ceiling materials may occasionally be installed on snap-in or other grid members that are directly clipped to structural elements or inserts in concrete slabs, but that installation is so similar to that described

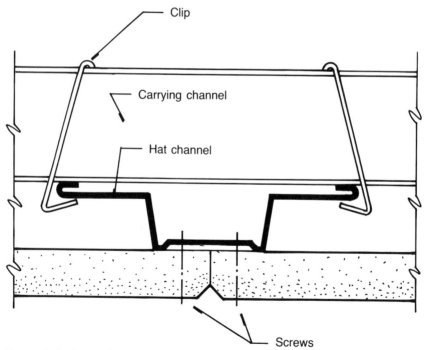

Figure 3-4   Acoustical tile directly applied to metal furring.

later in this chapter for a metal suspension system that the requirements addressed there apply.

Metal furring bars that directly support acoustical materials should comply with the building code and the standards and minimum requirements of generally recognized industry standards. They may be screwed or clipped to the supporting structure, or clipped or screwed to carrying channels, which are in turn suspended from the structure by wires or bars.

Hung Metal Suspension Systems. There are two categories of hung suspension systems for acoustical materials: direct-hung, in which the main runners are suspended from the structure, such as that shown in Figure 3-5; and indirect-hung, in which the main runners are suspended from carrying channels, such as that shown in Figure 3-6. The main runners in either category do not have to be those shown in Figures 3-5 or 3-6. They may be of any type that is appropriate for the acoustical material being supported. Some common types are shown in Figure 3-7, but there are many others.

Lay-in suspension systems may be either direct- or indirect-hung, depending on the system members and the necessary spacing of hangers. Concealed grid systems may also be either direct- or indirect-hung, but systems having splines or other less rigid cross members are usually indirect-hung. Except for the carrying channels, most members in both types of systems are proprietary, meaning that they are manufactured or furnished by a single supplier or producer for installation as a system.

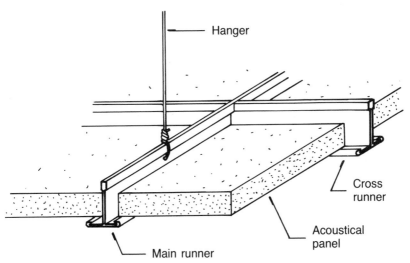

Figure 3-5   One type of direct-hung acoustical ceiling suspension system.

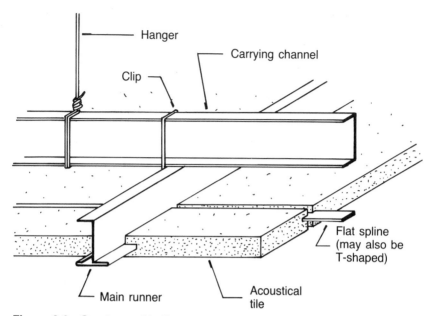

Figure 3-6   One type of indirect-hung acoustical ceiling suspension system.

Metal suspension systems for acoustical ceilings should comply with the applicable requirements of ASTM Standards C 635 and C 636. Acoustical treatment manufacturers may, however, have more stringent requirements.

Hung suspension systems for composition materials include concealed grid types for tile, and exposed or semi-exposed grid types for lay-in panels.

Hung suspension systems for metal acoustical ceiling materials include lay-in types and snap-in types. Their shapes vary with the manufacturer. Figure 3-8 shows a common T-bar snap-in type, but it is far from the only shape used. Some types of metal ceiling grids may be direct hung, even though they do not have cross-furring members because the metal ceiling material itself is stiff enough to stabilize the system without additional bracing.

Concealed Grid Systems for Composition Tile.   Most hung suspension systems designed to be concealed in the finished ceiling are H and T systems, which includes H-shaped main runners and T-shaped splines similar to those shown in Figure 3-7; concealed single- or double-web systems using T-shaped main runners similar to the standard tee shown in Figure 3-7, and spacer bars similar to those shown in the same figure at right angles to the main runners for stiffening; or Z systems (see Fig. 3-6), using Z's and splines similar to those shown in Figure 3-7. Other systems are also used

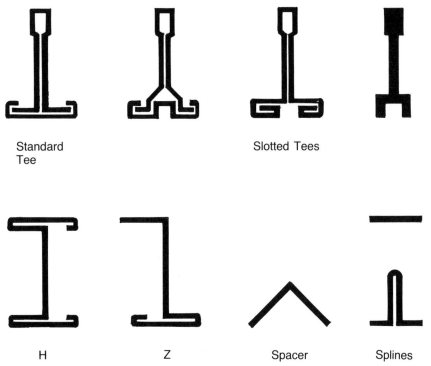

Standard Tee

Slotted Tees

H   Z   Spacer   Splines

Figure 3-7 Typical acoustical ceiling suspension system components.

today, and still others have been used in the past, so an existing installation may be different from those noted here.

A characteristic of all concealed suspension systems for composition tile is that they lock the tile in place so that it cannot be easily removed. To overcome this drawback, manufacturers have devised ways to make sections of tile hinged or removable, so that access can be gained where necessary. In direct-hung systems, hinged tile sections, after being released by a special tool, may be moved upward into the plenum or downward into the room. In both types, more than one tile is usually movable. Strips of tile between cross-framing members either lift up or can be pulled down to provide access. The movable panels of tile may be hinged along the cross runners or along the main runners. Once the movable tile has been moved out of the way, the remainder of the tile can be removed in successive pieces. Downward access systems are not recommended in some seismic zones.

In indirect-hung systems, access is sometimes gained with special cross runners or split splines that permit acoustical unit removal. Such special runners or splines are identified with unobtrusive permanent markers. Some

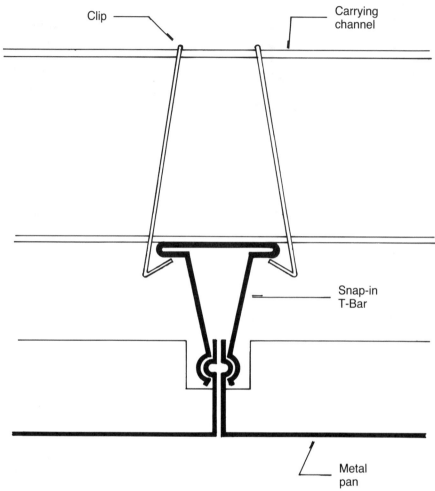

Figure 3-8 One type of T-bar snap-in metal ceiling system.

indirect-hung systems are made accessible with special tools that work only with that system.

Not all systems have a removable feature, however. Some older tile may be removable only by breaking out a tile and then removing the adjacent tile until the necessary opening has been achieved. Where the space above the ceiling is not accessible through the tiles, access doors must be provided.

Exposed and Semi-Exposed Grids for Composition Panels. Suspension systems designed for lay-in application of acoustical panels generally consist of T-shaped main runners usually called tees and interlocking cross tees (see Fig. 3-5). The cross members of semiconcealed systems are usually T-shaped splines. The main runners are usually directly hung from the structure with no carrying channels present. Main runner and cross runner shapes are proprietary, and vary with each system manufacturer. They are available with wide or narrow flat bottom surfaces, roll-formed caps, slotted, with reveals, and in other configurations. A few of the many configurations available are shown in Figure 3-7.

Grids for Metal Ceilings.   Metal ceilings are either of the lay-in or snap-in type. Some metal pans and panels lay into the same type grid used for composition material (see Fig. 3-5). Others hang from similar grids (Fig. 3-9).

Support grids for linear metal ceilings, while similar to that shown in Figure 3-10, vary from manufacturer to manufacturer to suit their particular ceiling material.

Many metal pans and some panels are installed into snap-in grids. The T-bar snap-in member shown in Figure 3-8 is only one of several snap-in device types in general use. Snap-in grid members are manufactured specifically to fit the edge configuration of the pans or panels being supported. A grid member made by one manufacturer may not correctly fit or support a panel made by a different manufacturer.

Structural Classification.   ASTM Standard C 635 divides hung suspension systems in three structural strength categories: light-duty, intermediate-duty, and heavy-duty. Light-duty systems, which are used in residential and light commercial installations, will support only the acoustical ceiling materials themselves. Intermediate-duty systems, which are the kind used in most commercial structures, will support light fixtures and air diffusers in addition to the acoustical materials themselves. Heavy-duty systems will support heavier loads than the other two classifications.

Edge Moldings and Trim.   With the exception of some linear metal ceilings, almost all acoustical ceiling suspension systems should have matching edge moldings at all vertical surfaces to which the acoustical ceiling abuts, including penetrations such as columns. Acoustical tile ceilings applied using fasteners or adhesives should also have such edge moldings and trim.

The edge molding and trim type and configuration should be suitable for the suspension system used. Some ceiling types have specialized trim members used only for that type.

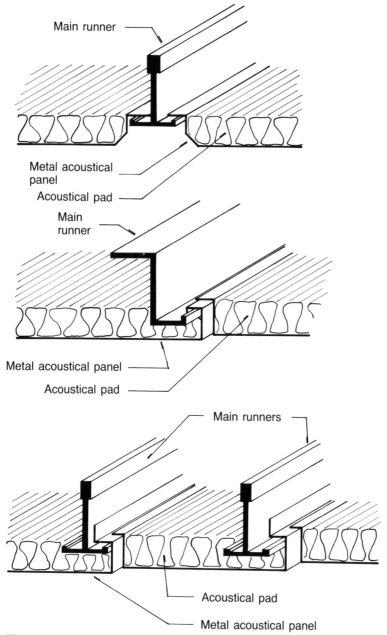

Figure 3-9 Some typical hung metal ceiling systems.

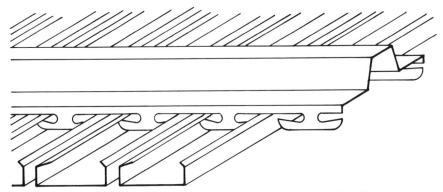

Figure 3-10 Typical support grid member for a linear metal ceiling.

Metals and Finishes. In exterior installations and in interior installations where high humidity will be present, such as in swimming pool areas or commercial kitchens, concealed metals should be either aluminum, stainless steel, zinc alloy, or hot-dip galvanized steel. The finishes on such metals should comply with ASTM Standard C 635 requirements, in the paragraph entitled "Coating Classification for Severe Environment Performance."

In locations other than those mentioned in the preceding paragraph, for composition tile and panels, the manufacturer's standard steel products are generally used with the manufacturer's standard galvanized or painted finish.

Exposed grids and edge members in exterior installations and in interior intallations where high humidity will be present, such as in swimming pool areas or commercial kitchens, should be either aluminum or stainless steel. Aluminum is usually protected by a baked-on paint, fluoropolymer, or anodized finish, but may also be left natural and protected by a coat of lacquer. Stainless steel may be protected by coating or left exposed. The finishes should comply with ASTM Standard C 635 requirements, in the paragraph entitled "Coating Classification for Severe Environment Performance."

Exposed grids and edge members in locations other than those mentioned in the preceding paragraph are generally steel finished with the manufacturer's standard paint finish, but may be aluminum or stainless steel where those decorative finishes are selected. Where the acoustical ceiling material has a decorative finish, such as anodized aluminum or stainless steel, exposed suspension system members often match the ceiling material and finish.

Colors of exposed finished metal vary, though some sort of white or off-white is probably the most common paint color.

Accessories and Miscellaneous Materials

The types of accessories needed for acoustical ceiling installation include those listed in the following paragraphs.

Carrying channels should be 1-1/2-inch cold-rolled steel channels, weighing a minimum of 0.475 pounds per linear foot, painted. Where spans require, 2-inch cold-rolled steel channels weighing a minimum 0.590 pounds per linear foot, painted, should be used.

Hanger wire for a typical installation should be galvanized carbon steel wire of a size necessary to properly support three times the hanger design loads, but in no case less than 12-gage wire. Some codes may require heavier wire regardless of loads. For the support of aluminum or stainless steel grid systems, and in exterior locations and air plenums, stainless steel or aluminum wire of a gage recommended by the suspension system manufacturer should be used.

Clips should be commercial types, acceptable under the building code for support.

Hold-down clips should be the suspension system manufacturer's standard types. Hold-down clips are necessary where uplift is a problem, whether from air pressure or to permit cleaning. Clips are also necessary in some fire-rated assemblies. Security clips are sometimes used to make metal pan ceilings nonremovable.

Tie wire should generally be galvanized wire and at least 16 gage.

Extension hangers for metal ceilings and hanger rods and flat bars should be zinc- or cadmium-coated mild steel. Alternatively, such hangers may be finished with rust-inhibitive paint.

Angle-shaped hangers should be formed from nominal 15- or 16-gage galvanized sheet steel. Legs should not be less than 7/8-inch wide.

Hanger anchorage devices should be rust-inhibitive screws, clips, bolts, concrete inserts, expansion anchors, or other devices. They should be sized to support five times the calculated hanger loading.

Screws and other fasteners should be rust-inhibitive and of a type standard with the suspension system manufacturer. Tile fasteners should be stainless steel or cadmium plated of the type recommended by the tile manufacturer, and should penetrate the substrate by at least 1/2 inch. Exposed fasteners should match the material fastened in appearance.

Acoustical sealant should be a water-based, nondrying, nonbleeding, nonstaining type, which is permanently elastic.

Most sound attenuation blankets are semirigid mineral fiber blankets with no membrane facing. They should have a Class 25 flame-spread. They should be 1-1/2 inches thick or more.

Acoustical pads for use in metal pan and panel ceilings are usually mineral or glass fiber. Some are wrapped or faced with materials such as polyvinyl chloride. Others are bare.

Adhesives for installing acoustical tile should comply with ASTM Standard D 1779 or Federal Specification MMM-A-00150, and should be UL labeled with a Class 0 to 25 flame spread rating.

Fire Performance

The Federal Specification SS-S-118 class (flame spread) rating refers only to the acoustical material itself. Underwriters Laboratories (UL) "Building Materials Directory" contains a listing showing the ratings of acoustical ceiling materials by manufacturer and product. Acoustical materials manufacturers usually indicate the flame spread rating of their products.

Hung suspension systems are classified as fire-resistance-rated or non-fire-resistance-rated, but sometimes it is hard to tell which is which from the manufacturers' product literature.

Unfortunately, using a rated acoustical material, suspension system, or both together does not alone result in compliance with code or other requirements for fire-rated construction. Fire-rated construction is a function of the entire assembly, including the floor or roof construction, flooring or roofing, insulation, and the installed ceiling. UL's "Fire Resistance Directory" includes a listing of construction systems, which includes the structure (floor, roof, etc.) and finishes, including acoustical ceilings. Ceiling materials are also listed there by manufacturer and product.

Fire-resistance ratings are established according to ASTM Standard E 119.

Installation

General Requirements. Layout plans should be prepared for spaces where acoustical ceilings will be installed to ensure coordination with lighting fixtures and other ceiling-mounted equipment and devices.

Manufacturer's brochures and installation details, and ASTM and CISCA standards, should be obtained and followed. Metal suspension systems should be installed in accord with ASTM Standard C 636, unless the acoustical treatment manufacturer has more stringent requirements.

The manufacturer's recommendations for cleaning and refinishing acoustical units and suspension system components should be obtained and stored in a safe place so that they will be available when future maintenance and repair are necessary. They should include precautions against materials and methods which may be detrimental to finishes and acoustical efficiency.

Extra material, in the amount of two or three percent of the amount installed, is often left at the site for the owner to use in making repairs. A search for this material is worthwhile when repairs are needed. Even when there is not enough to make the repairs, the stored material is often in the original cartons, which helps to identify the actual products used in the building.

Acoustical materials and suspension system components should be kept dry before and after use. Acoustical materials should be allowed to acclimate to the temperature and humidity of the space before they are installed to prevent dimensional changes, which might affect the appearance of the installed material.

Watermarked, rusted, or otherwise damaged units should not be installed.

Acoustical ceiling components should not be installed until the building is weathertight, glazing has been completed, exterior openings have been closed, and mechanical, electrical, fire-prevention (Fig. 3-11), and other work above ceilings has been completed.

Wet work, such as masonry, concrete, and plaster, should be allowed to dry to the moisture content recommended by the acoustical materials manufacturer before acoustical materials are installed. A temperature of at

Figure 3-11 Sprinklers are a common occurrence in acoustical ceilings. (*Photo by author.*)

least 60 degrees Fahrenheit should be maintained in the space during and after installation of acoustical materials.

Light acoustical panels in fire-rated assemblies generally require hold-down clips. Panels that require pressure-cleaning, such as those in commercial kitchens and hospital clean rooms and surgeries, also require hold-down clips.

Ceilings where cleanliness is paramount, such as in hospital operating rooms, often require sealant around the perimeter of the ceiling grid and sometimes even around the perimeter of each panel at its supports. Health department rules and regulations and building codes should be examined to determine the applicable requirements.

Suspension systems for use where fire-resistance ratings are required should comply with governing regulations, and their materials and installations should be identical with assemblies which have been tested and listed by recognized authorities, such as Underwriters Laboratories.

Suspension system members that abut vertical or overhead building structural elements should be isolated from structural movement sufficiently to prevent transfer of loads into the suspension system.

Provisions should be made in the suspension system to accommodate movement that occurs across control and expansion joints in the substrates.

Suspended acoustical ceilings should have a seismic restraint system in accordance with ASTM Standard E 580, where required by code or usual practice.

Adequate space should be left for sealants where they are required.

Flat surfaces of suspension system members should not be warped, bowed, or out of plumb or level by more than 1/8 inch in 12 feet in both directions, or by more than 1/16 inch across any joint or 1/8 inch total along any single member.

Before acoustical ceiling installation begins, the area should be thoroughly inspected to verify that mechanical, electrical, and other work above or within the acoustical ceilings has been completed. Acoustical ceiling work should not proceed until other work to be concealed or made inaccessible by the ceiling has been completed and unsatisfactory conditions have been corrected.

Acoustical units should be handled carefully to prevent damage to edges and faces.

Wood Furring Installation.   Wood furring should be installed as is appropriate to support the acoustical ceiling. It should be placed on lines and levels necessary to cause the acoustical material to fall into the proper location. It should correct unevenness in the supporting structure or substrate. The surface of wood furring to which acoustical material will be fastened should not be less than 1-1/2 inches wide. Generally, 2 by 2 lumber is used for furring that is attached to the supporting structure, and 1 by 2 or 1 by 3 lumber is used where furring is laid directly over solid substrates. Nominal 1-inch-thick lumber should not be used over framing, because it is too flexible. Other sizes may be found in existing construction, however, especially in older construction.

Sometimes acoustical ceilings are furred using 2 by 4's hung from structural floor or roof framing by hangers spaced not more than 48 inches on center. Frequently, in older buildings the hangers used were wood members instead of wire.

Furring should form a complete system adequate to properly support the acoustical tile. Closers should be installed at edges and openings.

Bolts, lag screws, and other anchors should be used to anchor the furring in place. Generally, fasteners are placed near the ends of furring members and not more than 36 inches on center between. Shorter members, however, should be anchored at 30 inches on center. Bolts should have nuts and washers.

Furring spacings may vary in practice with the substrate and the acoustical tile being supported. Minimum spacings are dictated by acoustical tile thickness and size, and may vary somewhat from manufacturer to manufacturer and system to system. Spacing must be such as to adequately support the acoustical tile.

Metal Furring Installation.   Metal furring should be installed as appropriate to support the acoustical material. It should be placed on lines and levels necessary to cause the acoustical material to fall into the proper location. It should correct unevenness in a supporting structure or solid substrate.

Furring should form a complete system adequate to properly support the acoustical material. Closers should be installed at edges and openings.

Screws, bolts, clips, and other anchors should be used to anchor the furring in place. Generally, fasteners should be placed near the ends of furring members and not more than 24 inches on center between, but other support spacings may be found. Some types of furring bars may require closer spacings. The correct spacing should be verified with the furring bar manufacturer.

Furring spacings may vary in practice with the substrate and the acoustical tile being supported. Minimum spacings are dictated by acoustical tile size and thickness, and may vary somewhat from manufacturer to manufacturer and system to system. Spacing must be such as to adequately support the acoustical tile.

Hung Suspension System Installation.   Suspension system hangers should be secured to wood, concrete, or steel structural supports by connecting directly to the structure where possible. Where direct connection is not possible, the hangers should be connected to inserts, clips, anchorage devices, or fasteners. Except for hanger attachments installed specifically for the purpose by the steel deck installer, hangers should not be attached to steel decks. Since failures have occurred even when deck tabs acceptable to the deck manufacturer have been used, some sources recommend that a ceiling not be supported from a steel deck under any circumstances. Hangers should never be attached to pipes, conduit, ducts, or mechanical or electrical devices.

Hangers should be located plumb in relation to primary suspension members. Contact with insulation covering ducts and pipes and other items within the plenum space should be avoided. Hangers should be splayed only where obstructions or conditions prevent plumb, vertical installation. Horizontal forces generated by splayed hangers should be offset by counter-splaying, bracing, or another suitable method.

Frames for grilles, registers, and other items that occur within acoustical ceilings should be properly installed in the correct locations.

Additional hangers should be provided for support of the grid systems at lights, grilles, registers, and other items to prevent excessive deflection. At grid systems with a light-duty structural classification, lights, grilles, registers, and other items should be supported separately from the ceiling grid.

Main runner and cross-furring members should be placed in proper positions, leveled, and fastened together in the manner required by the system. Tee runners are usually locked together without accessories. Some systems, however, require clips.

Except where the acoustical ceiling is self-edging, edging strips should be provided at the complete perimeter of all acoustical ceilings, including those with concealed grids, and where suspended ceilings abut vertical surfaces. Short lengths should be avoided. Standard moldings should be attached through the web 3 inches from each end, and not more than 16 inches on center between. Special moldings may require other spacings, as recommended by their manufacturers. Edging strips are usually of the same material and finish as exposed portions of grid systems.

Light fixtures and other penetrations are often boxed using mineral wool boards to comply with UL rating requirements where ceilings are rated.

Suspension systems should achieve a tight and level ceiling at the spacings and directions recommended by the manufacturer, and to comply with UL label requirements.

Installation of Concealed Grid Systems for Composition Acoustical Tile.   Main runners should be supported at spacings not greater than those recommended by the manufacturer. Maximum spacing of hangers is usually 48 or 60 inches, depending on the system and the components selected. The first hanger on each end of each runner or carrying channel should be no more than 6 inches from the ends. Some systems require closer spacings. Where the hangers must be placed at wider spacings than those recommended for the system, the main runners can be supported by carrying channels. In indirect-hung suspension systems, main runners are usually supported by carrying channels.

Installation of Exposed Grid Systems for Composition Acoustical Panels.   Tees should be provided of adequate size to span between supporting structure. If necessary, main tees may be supported from carrying channels.

Main tees are usually located 48 inches on center and all in the same direction. Cross tees are then installed to span between the main runners with cross tees at 24 inches on center. The tees should be locked together using whatever system is applicable to the system being used. The tees should rest on wall moldings at vertical surfaces. Where 24 inch square panels are used, additional cross tees should be placed from cross tee to cross tee and securely fastened (or locked) to the cross tees.

Main tees should be supported by hangers placed not more than 48 inches on center. Where the structure will not permit hangers at that spacing, additional carrying channels should be provided at 48 inches on center and the main runners should then be supported from the carrying channels. The first hanger on each end of each runner or carrying channel should be no more than 6 inches from the ends.

Exposed members should be installed in line, square, level, in one plane, and within the tolerances permitted by ASTM Standard C 636.

Composition Acoustical Tile Installation.   Where a grid system is used, tiles should be placed over the entire grid area. Where necessary, tiles should be cut to fit around fixtures, penetrating hangers, and accessories. Border units should be cut and fit neatly and snugly into the grid, leaving no voids in the finished ceiling.

Splines and other grid members should be placed into the kerfed edges of tiles or tile tongues should be inserted into tile grooves so that every joint is fitted tightly and properly supported. The tile field should be held in compression with leaf-type spring steel spacers between the tiles and the edge moldings. Spacers should be placed 12 inches apart.

When a grid system is not used, tiles should be held in place using fasteners, adhesive, or both, following the tile manufacturer's instructions. Before adhesive application, loose dust should be removed from the backs of the tiles and they should be primed with a thin coat of adhesive. Surfaces to receive tiles should be sound, free of foreign materials, clean, and dry.

New plaster and concrete to receive tiles should be cured thoroughly. Bare plaster, gypsum board, and concrete should be primed as recommended by the tile manufacturer. The alkalinity of surfaces to receive adhesive-applied tiles should be tested with litmus or Hydrian pH paper. Surfaces with a pH in excess of 9 should have a neutralizer applied to lower the pH to 8 or less.

Tiles are installed with adhesives by applying spots of adhesive to each tile and pressing the tiles in place using a sliding motion. The thickness of the adhesive after application should be about 1/8 inch.

Splines should be installed in the joints between tiles to keep them in proper alignment and flush with each other. Tiles should be scribed to fit at penetrations and edges.

When tiles are fastened in place using fasteners or adhesives, it is necessary to provide access doors in the ceiling where mechanical, electrical, or other equipment must be accessible. Such panels are usually recessed so that they can be faced with tiles like those installed in the ceiling.

Composition Acoustical Panel Installation.   Lay-in panels should be placed into each opening of the exposed grid members. Where necessary, panels should be cut neatly around fixtures, diffusers, penetrating hangers, accessories, and other items (Fig. 3-12). Border units should be cut to fit snugly into the grid, with no cracks or voids.

Metal Acoustical Materials Installation.   Linear metal ceiling strips should be snapped into the supporting grid members according to the ceiling manufacturer's recommendations, to cover the entire ceiling.

Lay-in metal panels and pans should be laid into each opening in the grid.

Snap-in panels and pans should be placed in each opening in the grid and shoved into place so that the holding feature fully engages and the panels or pans are held securely in place. Where a ceiling is to be non-accessible, security clips should be installed as the tile or pans are placed.

Sound Insulation.   Where they are used, sound insulation blankets may be installed over the full area occupied by the acoustical ceiling or only in certain locations, such as close to sound-rated partitions. Unless the ceiling is specifically designed to support sound insulation, the insulation should be separately supported so that its weight does not rest on the acoustical materials or any part of the suspension system.

Sound insulation pads in metal ceilings should fit snugly into each pan, panel, or strip over the entire ceiling area.

Insulation should be tightly butted with joints in solid contact and without voids. It should be fitted neatly to adjacent surfaces.

Figure 3-12   Unusual items sometimes occur in acoustical ceilings. (*Photo by author.*)

Acoustical Sealant.   Where light leaks, sound leaks, and air movement at the edge of acoustical ceilings is not acceptable, a continuous bead of acoustical sealant should be applied on the concealed leg of the ceiling edge molding before it is installed.

Why Acoustical Ceilings Fail

Some acoustical ceiling failures can be traced to one or several of the following sources: structure failure; structure movement; solid substrate problems; supported substrate problems; and other building element problems. Those sources are discussed in Chapter 2. Many of them, while perhaps not the most probable causes of acoustical ceiling failure, are more serious and costly to fix than the types of problems discussed in the following paragraphs. Consequently, the possibility that they are responsible for an acoustical ceiling failure should be investigated.

The causes discussed in Chapter 2 should be ruled out or, if found to be at fault, rectified before acoustical ceiling repairs are attempted. It will do no good to repair existing, or install new, acoustical ceilings when a failed, uncorrected, or unaccounted-for problem of the types discussed in

Chapter 2 exists. The new installation will also fail. After the problems discussed in Chapter 2 have been investigated and found to be not present, or repaired if found, the next step is to discover any additional causes for the failure and correct them. Possible additional causes include bad materials, improper design, bad workmanship, failing to protect the installation, and poor maintenance procedures. The following paragraphs discuss those additional causes. Included with each cause is a numbered list of errors and situations that can cause acoustical ceiling failure. Refer to "Evidence of Failure" later in this chapter for a listing of the types of failure to which the numbered failure causes apply.

Bad Materials. Improperly manufactured furring, suspension systems, or acoustical materials arriving at a construction site is certainly not unheard of, and that possibility should be considered when acoustical materials fail. Since most acoustical material imperfections can easily be seen, bad materials should be eliminated before they are installed. At any rate, the number of incidents of bad materials is small compared with the cases of bad design and workmanship.

Types of manufacturing defects that might occur include:

1. Acoustical materials that are not true to shape. Examples include square tiles or panels that are actually trapezoids, linear metal ceiling members which are not the same width throughout, and tile or panels that are warped.
2. Acoustical materials with inconsistent surface patterns.
3. Acoustical materials that are stained or otherwise discolored at the factory, or are manufactured using materials that are inconsistent in color.
4. Acoustical materials that are damaged in the manufacturing process.
5. Acoustical materials that are inconsistent in composition or density.
6. Furring or suspension system members that are twisted, bent, bowed, warped, or otherwise malformed, or which do not have sufficient camber.

Improper Design. Improper design includes selecting the wrong furring or suspension system or acoustical material for the location and conditions, as well as requiring an improper installation.

General Problems. The following general design problems can lead to acoustical ceiling failure:

1. Selecting acoustical materials with inappropriate faces or composition for the location and use intended. Units with standard painted faces are not generally suitable for use over swimming pools, for ex-

ample. Units with surfaces that are not abuse-resistant are not suitable for use where they can be touched by people.

2. Selecting acoustical material that has an inappropriate acoustical performance rating. Using material that has not been tested for speech privacy in an open-plan office is one example. Selecting material with too low an STC or NRC rating for the noise levels that will occur is another.

3. Failing to provide a means to gain access to the space above an acoustical ceiling.

4. Requiring that loads, such as those from blanket insulation, electrical or mechanical system components, light fixtures, sound system components, or other items, be supported by a suspension system with too low a structural classification to permit such installation.

5. Requiring that loads, such as those from blanket insulation, electrical or mechanical system components, light fixtures, sound system components, or other items, be supported by acoustical materials. Metal pans and panels, of course, are usually designed to receive acoustical pads. Linear metal ceilings are sometimes designed to support sound insulation blankets.

6. Selecting fasteners, accessories, hangers, suspension system grid components, or other materials that will rust, for use where high humidity will occur.

7. Selecting fasteners, clips, and other accessories that are of the wrong type for the installation, or are too small or weak to perform the necessary function.

8. Failing to require proper installation of acoustical tile using fasteners or adhesives, including requiring that too few fasteners or too little adhesive be used, that fasteners be placed in the wrong locations, that adhesive be improperly applied to the tile, or that tile be improperly installed after the adhesive has been applied.

9. Requiring too many light fixtures or other relatively hard surfaces in the ceiling. Speech privacy may be impaired if more than ten percent of the ceiling surface is covered with flat-lens recessed light fixtures.

10. Locating light fixtures or other hard-surfaced items directly over a space divider, thus permitting sound to reflect off the hard surface from one area into an adjacent area.

Wood Furring Problems. Wood furring may fail because of structure failure, structure movement, solid or supported substrate failure, or other building element problems, as discussed in Chapter 2, or because of problems inherent in the furring itself.

Furring design problems that can cause acoustical treatment failures include the following:

1. Requiring the use of flexible or extremely hard wood for furring.
2. Permitting the use of wet lumber.
3. Requiring that furring members be placed too far apart.
4. Requiring that wood furring be installed so that loads from a deflecting or otherwise moving structure pass into the furring.

Metal Furring Problems.   Metal furring may fail because of structure failure, structure movement, solid or supported substrate failure, or other building element problems, as discussed in Chapter 2, or because of problems inherent in the furring itself.

Furring design problems that can cause acoustical treatment failures include the following:

1. Requiring that furring members be placed too far apart.
2. Requiring that metal furring be installed so that loads from a deflecting or otherwise moving structure pass into the furring.

Metal Suspension System Problems.   Metal suspension systems may fail because of structure failure, structure movement, solid substrate problems, or other building element problems, as discussed in Chapter 2. In addition, a suspension system may fail because of problems inherent in the suspension system itself.

Metal suspension system design problems that can cause acoustical ceiling failures include the following:

1. Selecting a system with too weak a structural classification for the conditions.
2. Requiring that suspension system hangers be placed too far apart or that hangers be too small for the loads to be supported.
3. Requiring that suspension system hangers be attached to inappropriate supports, such as pipes, conduit, ducts, or steel decks.
4. Requiring use of the wrong material for hangers. Requiring use of annealed wire over a swimming pool, for example.
5. Requiring use of inappropriate methods of tying hanger wire.
6. Requiring that metal suspension system components be installed so that loads from a moving structure will pass into the suspension system.

Bad Workmanship.   Correct installation is essential if acoustical ceiling failures are to be prevented.

General Problems.   The following workmanship problems can lead to acoustical ceiling failure.

1. Failing to follow the design and the recommendations of the manufacturer and recognized authorities, such as ASTM.
2. Failing to properly prepare the area where an acoustical ceiling will be installed, removing contaminants and items that will interfere with proper installation or damage the ceiling materials.
3. Failing to properly clean the back of acoustical tile that will be installed using adhesives and to prime the substrate before tile installation.
4. Installing acoustical ceiling components before the building has been completely closed and wet work, such as concrete, masonry, and plaster, have dried out sufficiently and, where tile is to be installed using adhesives, are at the proper level of alkalinity.
5. Installing damaged acoustical materials, furring, or suspension system components, regardless of whether the damage was inherent in the manufactured materials or occurred during shipment, storage, or installation.
6. Using fasteners, hangers, accessories, and other materials that will rust where high humidity will occur or where set in, or in direct contact with, masonry or concrete.
7. Using fasteners, clips, wire, and other accessories that are of the wrong type for the installation, or are too small or weak to perform the necessary function.
8. Installing acoustical ceilings so that loads, such as those from blanket insulation, electrical or mechanical system components, light fixtures, sound system components, or other items, are supported by a suspension system with too low a structural classification to permit such installation.
9. Failing to allow acoustical materials to acclimate to the temperature and humidity in the space before installing them, which can lead to appearance changes in the installed ceiling.
10. Permitting acoustical materials to become wet or damp before or during installation, which can cause discolorations due to water marks (Fig. 3-13) and mildew, or otherwise damage composition materials or contribute to the oxidization of metal materials.
11. Installing acoustical materials that are watermarked or otherwise stained.
12. Installing acoustical materials that have damaged edges or faces (Fig. 3-14).
13. Permitting loads, such as those from blanket insulation, electrical or mechanical system components, light fixtures, sound system

Figure 3-13   Watermarked acoustical panel. (*Photo by author.*)

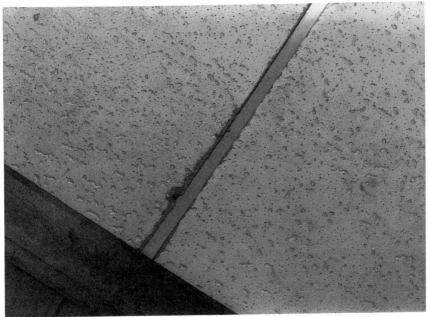

Figure 3-14   Acoustical panel with damaged edges. (*Photo by author.*)

63

components, or other items, to fall on acoustical materials that are not designed to support them.

14. Failing to install hold-down clips where required.
15. Failing to insert splines in the joints of acoustical tiles installed using fasteners or adhesives.
16. Failing to install spring steel spacers at the perimeter to hold acoustical tile tightly in place.
17. Failing to properly install acoustical tile using fasteners or adhesives, including not using enough fasteners or adhesive, placing fasteners in the wrong locations, improperly applying adhesive to the tile, and improperly installing the tile after the adhesive has been applied to it.
18. Failing to install acoustical ceiling components so that the ceiling finishes in the proper plane within acceptable tolerances.
19. Failing to seal around the perimeter of an acoustical ceiling.
20. Failing to install a complete system, such as leaving out a tile.
21. Installing acoustical tile using adhesives when the room temperature exceeds 100 degrees Fahrenheit, or when the room or adhesive temperature is less than 50 degrees Fahrenheit.
22. Applying acoustical tile, using an adhesive, to damp plaster or concrete, or in a damp room.
23. Applying acoustical tile to loose paint using an adhesive.
24. Failing to install acoustical materials in or over metal ceilings.

Wood Furring Problems. Acoustical ceilings may fail because of problems inherent in wood furring, including:

1. Installing furring so that it is misaligned, twisted, or protruding beyond the plane it should be in.
2. Lumber shrinkage. Even relatively dry wood will shrink.
3. Using wet lumber. When it dries, lumber that is too wet when installed can warp, split, twist, or otherwise change in shape, or may shrink away from fasteners and become loose or fall away from its supports.
4. Using flexible or extremely hard wood for furring. Wood that is too hard or too flexible may prevent fasteners from entering and seating properly, resulting in loose acoustical tile.
5. Using wood that is too thin for its span.
6. Installing furring members too far apart.
7. Failing to properly support and anchor furring members. When the furring loses stability or separates from the substrates, acoustical treatment failure is inevitable.

8. Leaving out portions of the furring, such as at openings and edges, so that there is nothing to which the acoustical tile may be fastened.

9. Installing wood furring so that loads from a deflecting or otherwise moving structure pass into the furring. Examples include wedging the ends of furring against masonry or concrete and building the ends of furring members into masonry.

Metal Furring Problems.   Acoustical ceilings may fail because of problems inherent in metal furring, including the following:

1. Installing furring so that it is misaligned, twisted, or protruding beyond the plane it should be in.

2. Placing furring members too far apart.

3. Failing to properly support and anchor furring members. When the furring loses stability or separates from the substrates, acoustical treatment failure is inevitable.

4. Leaving out portions of the furring, such as at openings and edges, so that there is nothing to which the acoustical tile may be fastened.

5. Installing metal furring so that loads from a deflecting or otherwise moving structure pass into the furring. Examples include wedging the ends of furring against masonry or concrete and building the ends of furring members into masonry.

Metal Suspension System Problems.   Acoustical ceilings may fail because of problems inherent in suspension systems, including the following:

1. Placing suspension system components so that they do not properly align with each other or adjacent walls.

2. Installing twisted or bent suspension system grid members. Members may be improperly manufactured or damaged during shipment, storage, or handling.

3. Placing suspension system components at an incorrect spacing, either foo far apart or too close together. This can cause acoustical materials to fit improperly or fall out.

4. Installing suspension system hangers too far apart or with undersized wire.

5. Fastening hangers to steel decks, pipes, conduit, ducts, or mechanical or electrical devices.

6. Splaying hangers without compensating for the forces generated by doing so.

7. Making wire ties incorrectly or poorly.

8. Using the wrong wire hanger material, such as using annealed wire over a swimming pool.

9. Failing to provide additional hangers for a grid system at lights, grilles, registers, and other items, where the grid system is not designed to support such items. Failure to separately support insulation and other devices where the structural classification of a ceiling will not permit their support by the ceiling system.

10. Failing to properly lock or attach suspension system grid components together. When the suspension system grid components lose stability or separate from each other, acoustical ceiling failure is inevitable.

11. Installing metal suspension system components so that loads from a moving structure will pass into the suspension system. Ceiling suspension system components should be held away from adjacent solid substrates and structural elements.

12. Failing to properly attach wall moldings and trim in place.

Failing to Protect the Installation.   Acoustical ceiling components must be protected before, during, and after installation. Errors include the following:

1. Failing to protect acoustical ceilings from staining or other damage by other construction materials or procedures.

2. Permitting abuse before or during installation, or after the acoustical ceiling has been installed.

Poor Maintenance Procedures.   Not maintaining acoustical ceilings properly can result in failures. Poor maintenance procedures include:

1. Failing to clean ceilings regularly. Unremoved grease, dirt, and other contaminants can result in permanent stains.

2. Improper cleaning of acoustical materials. Hard finishes, such as those on metal acoustical ceiling materials, and even those designed for exposure to wet environments or for use in areas where people can reach them, will probably be damaged by abrasive cleaners. Standard acoustical material finishes, which may be harmed even by contact with water, may be damaged severely by abrasive cleaners or even detergents.

3. Improper cleaning of exposed suspension system components. Abrasive cleaners or rough treatment can damage finishes.

4. Improperly painting acoustical materials.

Evidence of Failure

In the following paragraphs, acoustical ceiling failures are divided into failure types, such as "Stained or Discolored Acoustical Materials." Following each failure type are one or more failure sources, such as "Improper Design:

General Problems." After each failure source, one or more numbers is listed. The numbers represent possible errors associated with that failure source that might cause that failure type to occur.

A description and discussion of the numbered failure causes for failure types "Steel and Concrete Structure Failure," "Wood Sructure Failure," and "Structure Movement" appear under those headings in Chapter 2, under the main heading "Structural Framing Systems," and are listed in the Contents.

A description and discussion of the numbered failure causes for failure types "Solid Substrate Problems" and "Supported Substrate Problems" appear under those headings in Chapter 2, under the main heading "Substrates," and are listed in the Contents.

A description and discussion of the numbered failure causes for failure type "Other Building Element Problems," which is listed in the Contents, appears under that heading in Chapter 2, under the main heading "Other Building Elements."

The following example applies to the failure types listed in the preceding five paragraphs. Clarification and explanation of the numbered cause (1) in the example

▪ Supported Substrate Problems: 1 (see Chapter 2).

appears in Chapter 2 under the main heading "Substrates," subheading "Supported Substrate Problems," Cause 1, which reads

1. Loose Plaster.

A description and discussion of the types of problems and numbered failure causes that follow failure types "Bad Materials," "Improper Design," "Bad Workmanship," "Failing to Protect the Installation," and "Poor Maintenance Procedures" appears under those headings earlier in this chapter under the main heading "Acoustical Ceilings" subheading "Why Acoustical Ceilings Fail," which is listed in the Contents. Clarification and explanation of the type of problem (Metal Suspension System Problems) and numbered cause (1) in the example

▪ Improper Design: Metal Suspension System Problems: 1.

appears earlier in this chapter under the main heading "Acoustical Ceilings" subheading "Why Acoustical Ceilings Fail," which are listed in the Contents, sub-subheading "Improper Design," sub-sub-subheading "Metal Suspension System Problems," Cause 1, which reads:

1. Selecting a system with too weak a structural classification for the conditions.

The following are possible acoustical ceiling failure types.

Poor Acoustical Performance.   Every acoustical ceiling failure type listed in this chapter can affect acoustical performance. In addition, ceilings may exhibit unsatisfactory acoustical performance due to the following causes.

- Other Building Element Problems: 1, 2 (see Chapter 2).
- Bad Materials: 5.
- Improper Design: General Problems: 2, 9, 10.
- Bad Workmanship: General Problems: 1, 4, 5, 10, 12, 19, 20, 24.
- Poor Maintenance Procedures: 4.

Stained or Discolored Acoustical Materials.   Blemishes, including watermarks, mildew, and other stains and discolorations, may occur due to one or more of the following causes.

- Solid Substrate Problems: 1 (see Chapter 2).
- Supported Substrate Problems: 5, 6, 13 (see Chapter 2).
- Other Building Element Problems: 1, 2 (see Chapter 2).
- Bad Materials: 3, 4.
- Improper Design: General Problems: 1, 6.
- Improper Design: Wood Furring Problems: 2.
- Bad Workmanship: General Problems: 1, 2, 4, 5, 6, 10, 11.
- Bad Workmanship: Wood Furring Problems: 3.
- Failing to Protect the Installation: 1, 2.
- Poor Maintenance Procedures: 1, 2.

Damaged Acoustical Materials.   Damage may be cracked or split units, broken edges, depressions in unit faces, or holes through units. Any one or several of the following may be the cause, depending on the type and extent of the damage.

- Steel and Concrete Structure Failure: 1 (see Chapter 2).
- Wood Structure Failure: 1, 2, 3, 5, 6, 8 (see Chapter 2).
- Structure Movement: 1, 2, 3, 4, 5, 7, 8 (see Chapter 2).
- Solid Substrate Problems: 2, 3, 4 (see Chapter 2).
- Supported Substrate Problems: 3, 9, 11, 12 (see Chapter 2).
- Other Building Element Problems: 3, 4 (see Chapter 2).
- Bad Materials: 2, 4.
- Improper Design: General Problems: 1, 3, 5.
- Improper Design: Wood Furring Problems: 4.
- Improper Design: Metal Furring Problems: 2.

- Bad Workmanship: General Problems: 1, 2, 5, 9, 12, 13.
- Bad Workmanship: Wood Furring Problems: 1, 2, 3, 7, 9.
- Bad Workmanship: Metal Furring Problems: 1, 3, 5.
- Bad Workmanship: Metal Suspension System Problems: 2, 3, 6.
- Failing to Protect the Installation: 1, 2.
- Poor Maintenance Procedures: 2.

Loose, Sagging, Fallen, Raised, or Out-of-Line Acoustical Materials. Acoustical materials may become loose, sag, fall out of, or raise up in, a suspension system, separate from a substrate or furring, or be out of alignment with other units, due to one or more of the following causes.

- Steel and Concrete Structure Failure: 1 (see Chapter 2).
- Wood Structure Failure: 1, 2, 3, 4, 5, 6, 7, 8 (see Chapter 2).
- Structure Movement: 1, 2, 3, 4, 5, 6, 7, 8 (see Chapter 2).
- Solid Substrate Problems: 1, 2, 3, 4 (see Chapter 2).
- Supported Substrate Problems: 1, 2, 3, 4, 8, 9, 10, 11, 12, 14, 15, 16 (see Chapter 2).
- Other Building Element Problems: 1, 2, 3 (see Chapter 2).
- Bad Materials: 1, 6.
- Improper Design: General Problems: 1, 5, 7, 8.
- Improper Design: Wood Furring Problems: 1, 2, 3, 4.
- Improper Design: Metal Furring Problems: 1, 2.
- Bad Workmanship: General Problems: 1, 2, 3, 4, 5, 7, 9, 10, 13, 14, 15, 16, 17, 18, 21, 22, 23.
- Bad Workmanship: Wood Furring Problems: 1, 2, 3, 4, 5, 6, 7, 8, 9.
- Bad Workmanship: Metal Furring Problems: 1, 2, 3, 4, 5.
- Bad Workmanship: Metal Suspension System Problems: 1, 2, 3, 10.
- Failing to Protect the Installation: 1, 2.

Damaged Suspension System. Damage may include scratches; rust; or bent, twisted, or bowed members, and may result from one or more of the following causes.

- Steel and Concrete Structure Failure: 1 (see Chapter 2).
- Wood Structure Failure: 1, 8 (see Chapter 2).
- Structure Movement: 1, 2, 3, 4, 5, 6, 7 (see Chapter 2).
- Other Building Element Problems: 1, 2, 3, 4 (see Chapter 2).
- Bad Materials: 6.
- Improper Design: General Problems: 3, 4, 6, 7.

- Improper Design: Metal Suspension System Problems: 1, 2, 6.
- Bad Workmanship: General Problems: 1, 4, 5, 6, 7, 8.
- Bad Workmanship: Metal Suspension System Problems: 2, 6, 11, 12.
- Failing to Protect the Installation: 2.
- Poor Maintenance Procedures: 3.

Sagging, Fallen, or Out-of-line Ceiling Suspension Systems.   Suspension systems may sag, fall, or be out of alignment with each other or other building elements, due to one or more of the following causes.

- Steel and Concrete Structure Failure: 1 (see Chapter 2).
- Wood Structure Failure: 1, 8 (see Chapter 2).
- Structure Movement: 1, 2, 3, 4, 5, 7, 8 (see Chapter 2).
- Other Building Element Problems: 1, 2, 3, 4 (see Chapter 2).
- Bad Materials: 6.
- Improper Design: General Problems: 4, 6, 7.
- Improper Design: Metal Suspension System Problems: 1, 2, 3, 4, 5, 6.
- Bad Workmanship: General Problems: 1, 5, 6, 7, 8.
- Bad Workmanship: Metal Suspension System Problems: 3, 4, 5, 6, 7, 8, 9, 10, 11, 12.
- Failing to Protect the Installation: 2.

Cleaning, Repairing, Refinishing, and Extending Acoustical Ceilings

General Requirements.   The following paragraphs contain some suggestions for cleaning, repairing, refinishing, and extending acoustical ceiling materials and associated furring and suspension systems. Because the suggestions are meant to apply to many situations, they might not apply to a specific case. In addition, there are many possible cases that are not specifically covered here. When a condition arises in the field that is not addressed here, advice should be sought from the additional data sources mentioned in this book. Often, consultation with the manufacturer of the materials being repaired will help. Sometimes, it is necessary to obtain professional help (see Chapter 1). Under no circumstances should the specific recommendations in this chapter be followed without careful investigation and application of professional expertise and judgment.

Before an attempt is made to clean, repair, refinish, or extend existing acoustical ceilings, the existing materials manufacturers' brochures for products, installation details, and recommendations for cleaning and refinishing should be available and referenced. It is necessary to be sure that the manufacturers' recommended precautions against materials and methods which may be detrimental to finishes and acoustical efficiency are followed.

Where existing acoustical materials to be repaired or extended are installed using staples, nails, screws, or other fasteners, or are installed using adhesives, their manufacturer's recommendations for methods to be used in the repairs or extensions should be obtained. Often, it is not possible to satisfactorily remove materials installed in such ways without damaging the units to the extent that they are not suitable for reuse. Trying to do so without following the manufacturer's instructions will almost certainly lead to failure.

A check should be made to see whether sufficient extra material was left from the original installation to make the repairs. Often, as much as three percent of the amount of acoustical material actually installed, or two full cartons of each type, will be left with the owner for making future repairs.

Unless existing acoustical ceilings must be removed to facilitate repairs to other portions of the building, or are scheduled to be replaced with new materials, only sufficient materials should be removed as is necessary to affect the cleaning and repairs. The more materials that are removed, the greater the possibility of additional damage occurring during handling and storage of the removed materials. Materials that are removed should include those that are loose, sagging, or damaged beyond in-place repair. Other existing panels, tile, and furring, or suspension components that are sound, adequate, and suitable for reuse may be left in place, or removed, cleaned, and stored for reuse.

After acoustical materials have been removed, concealed damage to furring or suspension system components may become apparent. Such concealed damage should be repaired before repaired, cleaned, or new acoustical materials are installed. Such additional damage may include furring or suspension system components that are damaged, rusted beyond repair, or otherwise unsuitable for reuse. Existing furring, hangers, hanger attachments, and suspension system grid components that are sound, adequate, and suitable for reuse may be left in place, or removed, cleaned, and stored for reuse.

Existing acoustical materials, furring, and suspension system components that are to be removed should be removed carefully, and adjacent surfaces should be protected, so that the process does no damage to the surrounding area. Materials and components should be removed in small sections. Where necessary to avoid collapse, temporary supports should be installed to prevent the collapse of construction that is not to be removed. Removed materials that will be reinstalled should be handled carefully, stored safely, and protected from damage. Debris should be removed promptly so that it will not be responsible for damage to materials that are to remain in place.

Unless the decision is made to discard them, damaged acoustical materials, furring, and suspension system components that can be satisfactorily repaired should be repaired, whether they have been removed or left in place. A failure to repair known damage may lead to additional failure later on. Methods recommended by the manufacturer of the materials should be followed carefully when making repairs, even when the repairs are as simple as touch-up painting. Acoustical materials, furring, and suspension system components that cannot be satisfactorily repaired should be discarded and new matching units installed.

Areas where repairs will be made should be inspected carefully to verify that existing components that should be removed have been removed, and that the substrates and structure are as expected and are not damaged. Sometimes, substrate or structure materials, systems, or conditions are encountered which differ considerably from those expected, or unexpected damage is discovered. Both damage that was previously known and damage found later should be repaired before acoustical ceiling components are reinstalled. Acoustical ceiling reinstallation should not proceed until other work to be concealed or made inaccessible by the ceiling has been completed and unsatisfactory conditions have been corrected.

With a few exceptions, removed acoustical materials, furring, and suspension system components that are in an acceptable condition may be reinstalled. Where possible, they should be reinstalled in the same location from where they were removed. Materials and components that are installed in locations other than those from which they were removed may not match the surrounding materials, which will make patches obvious. An exception, of course, would be installing a removed ceiling in another space where the removed ceiling system is sufficient in quantity to completely fill the ceiling area in the new space.

Removed hangers and hanger attachments should not be reused unless they are in exceptionally good condition. Metal fatigue or other difficult-to-detect damage may exist, which may become apparent only when the reused hanger or hanger attachment fails.

It will probably not be possible to reinstall acoustical materials that were installed using staples, screws, nails, or other similar fasteners. The removing process will usually damage them too severely.

Unless a decision is made to leave a substrate exposed that was exposed by removing all or part of an existing acoustical ceiling, or to install another ceiling type or a new acoustical ceiling, salvaged acoustical materials should be reinstalled in accordance with the requirements of the referenced standards, such as ASTM C 635 and C 636, and the manufacturer's recommendations.

Where acceptable existing materials are insufficient in quantity, new matching materials should be used to complete the ceiling. During the process, gaps and openings in the existing ceiling, including both those that existed before the repairs began and those caused by discarding unrepairable materials, should be filled with new acoustical material that matches the existing materials.

Where patching or extending an existing acoustical ceiling is required, and acoustical materials or suspension systems do not exist, or existing materials are not acceptable for reuse, or existing materials are in insufficient quantity to complete the work, new acoustical materials and suspension systems should be provided. The new acoustical materials and suspension systems should match the existing adjacent undisturbed (or originally installed and removed) material in type, size, material, edge condition, thickness, pattern, finish, and characteristics. The new suspension system, or installation method, as applicable, should exactly match, as applicable, that existing in similar installations, or used in the original removed ceilings, or still existing in adjacent ceilings. Substrates should be repaired and cleaned as necessary to obtain a satisfactory installation.

Patched or extended work should be complete with no voids or openings and should produce a finished ceiling in the same plane as the existing ceiling left in place. Joints should be of the same type as those in the existing ceiling in every respect, and should be aligned with the existing joints. Patches and extensions should be made as inconspicuous as possible.

Damaged acoustical materials and suspension systems should not be reinstalled. Minor damage may be repaired, however, and the repaired components installed if satisfactory results can be obtained.

Removed edging strips should be reinstalled at the complete perimeter of suspended acoustical ceilings, including those with a concealed grid, and where suspended ceilings abut vertical surfaces. Short lengths should be avoided. Moldings should be attached through the web 3 inches from each end and not more than 16 inches on center between. Edging strips should be of the same material and finish as the exposed portions of the grid systems.

Reinstalled exposed members should be in line, square, level, in one plane, and within tolerances permitted by ASTM C 636.

Care should be exercised to ensure that the integrity of fire-rated systems is maintained. Mineral wool board boxes at lighting fixtures and other penetrations should not be altered or destroyed where they are required to comply with UL rating requirements. Acoustical materials, framing, furring, and suspension system materials and installation should be identical with applicable assemblies that have been tested and listed by recognized authorities and are in compliance with the requirements of the building code. Materials for use in existing fire-rated assemblies should, unless doing so violates the previous sentence, exactly match the materials in the existing fire-rated assembly.

Wood Framing and Furring. Repair of major wood framing members such as joists, rafters, trusses, truss-joists, beams, and girders is beyond the scope of this book. The following paragraphs contain only enough data to help in recognizing that those elements may be at fault when acoustical ceilings fail, and to present a general indication of what steps might be necessary to solve the problem.

Most failed wood framing or furring members are removed, and new material installed, because (even when doing so is possible) it is not often reasonable to repair damaged wood framing or furring members. Straightening a warped or twisted joist, for example, will probably prove to be impracticable. So, most of the references in the following paragraphs to repairing framing or furring mean removing the damaged pieces and installing new pieces.

The following paragraphs assume that damaged concrete or metal structure and solid or supported substrates supporting wood furring has been repaired and presents a satisfactory support system for the furring being repaired.

Materials.   Materials used to repair existing wood framing or furring should match those in place as nearly as possible, but should not be lesser in quality, size, or type than those recommended by recognized authorities or required by the building code. Where the existing materials are fire-retardant treated, the new materials must be similarly treated. Where the existing material is pressure preservative treated, the new should be also.

It is usually best to match lumber sizes exactly when installing a new member in an existing framing or furring system. In older installations, however, the existing lumber may be of sizes that are no longer standard. Members that are nominally 2 by 4 may be actually 2 by 4 inches if the building is old, 1-5/8 by 3-5/8 inches, as was once the standard, or 1-1/2 by 3-1/2 inches, as is the current standard. When lumber of the exact size to match that existing is not available, there are several possible ways to solve the problem. For example, larger members can be cut down to the size of the existing members. This may be an expensive solution, however. For furring, it will usually be less expensive to shim the new members so that the face surfaces are in alignment with the faces of the existing members. Another alternative is to build up new members from two standard-sized members.

Standards.   In general, repairs should be made in accordance with the recommendations of recognized standards, such as those mentioned earlier in this chapter or in "Where to Get More Information" at the end of this chapter, or both, and the standards referenced in the Bibliography as applicable to this chapter.

Preparation.   Where existing framing or furring members are damaged, the acoustical treatment must be removed to the extent necessary to permit the repairs to be made. Then the damaged existing framing or furring can be removed, along with related suspension systems, if any. Damaged elements that should be removed include wood members that are twisted, warped, broken, rotted, wet, out of alignment, or otherwise unsuitable for use. Existing hangers that are sound, adequate, and suitable for reuse may be left in place, or removed, cleaned, and reused.

Hangers and hanger attachments that have been left in place where suspended ceiling furring has been removed should be examined to verify their adequacy and suitability to support the new furring. Those found to be unsuitable should be removed.

Existing hangers and hanger attachments in good condition may be used to support new ceiling furring. New hangers and hanger attachments should be provided, of course, where existing hangers and attachments have been removed, where existing hangers or attachments are improperly placed or otherwise inappropriate, and where there are no existing hangers. New hangers should be attached to the structure, never to pipes, conduits, ducts, or mechanical or electrical devices. Existing hangers so attached should not be used.

Misaligned, Warped, or Twisted Framing or Furring.   Even when so misaligned, twisted, or warped that it causes damage to applied or suspended acoustical treatment, wood framing or furring is often left in place, and the acoustical ceiling reattached in such a way that the poor condition of the framing or furring is overcome. When repair to the acoustical treatment alone will not prevent failure from recurring, it will be necessary to remove the damaged framing or furring and provide new materials.

Shrinkage.   Where acoustical ceiling failure is caused by shrinkage in the lumber used in framing or furring, it is often possible to repair and reattach the acoustical treatment without removing the lumber. Where acoustical treatment repairs cannot be accomplished satisfactorily, it may be possible to remove a portion of the acoustical treatment and shim the framing or furring to produce a flush surface for acoustical ceiling application. Where shimming does not solve the problem, it may be necessary to remove the damaged lumber and provide new framing or furring.

Incorrectly Constructed Framing or Furring.   Where the framing or furring was originally built in such a way that the acoustical ceiling becomes damaged, the necessary corrective measures depend on the type of error. The possibilities are many, and the solutions even more numerous. They range from simply planing down a projecting brace to reconstructing a section of floor framing.

Each case must be examined to determine the true cause before steps are taken to correct supposed errors. The extent of the damage must be also taken into account. If it might be possible to prevent the failure from recurring by simply refastening the acoustical ceiling, for example, that should be done before expensive reconstruction of the framing or furring is undertaken.

Failed Ceiling Furring.   When an acoustical ceiling that is directly applied to wood furring sags or bows, the cause is often due to failure of the furring or its suspension system.

Wood ceiling support systems fail for several reasons, including loads placed on the ceiling furring from above, such as storage, people crawling around above the ceiling, structure deflection, differential movement in the structure, improper attachment of the support system, or other trauma. Failure includes broken wire hangers; broken wood hangers; and wood hangers that have dried out, shrunk, and split away from the nails fastening the ceiling furring nailers to the hangers.

When an acoustical ceiling fails because of a suspended furring system failure, repairs may be made by lifting the ceiling back into place and repairing the failed support system. The first step is to stabilize the ceiling. To do so, access to the space above the ceiling must be obtained. When access is not otherwise possible, it will be necessary to remove portions of the acoustical material. If the acoustical ceiling has been applied using fasteners or adhesives, it will probably be necessary to discard the removed acoustical material and provide new material.

After obtaining access to the space above the ceiling, the next step is to wedge the ceiling up from below into proper alignment. Wedging is often accomplished using T-shaped wood braces.

When the ceiling is once again in its proper position, lag bolts can be driven into the supporting structure above and the horizontal ceiling furring members wire-tied to the lag bolts.

When an acoustical ceiling failure is due to directly attached wood furring that has not been properly attached or where the furring attachment has failed, repairs may be made by lifting the ceiling into its proper position and refastening the furring to its supports. Of course, if the acoustical material is not removed while the refastening is being done, the new fasteners will probably be visible. If the furring is supported on metal, driving new fasteners through the ceiling will probably not be possible. In the latter two cases, removal of the acoustical material will probably be necessary. Reinstallation may be impracticable, however, if the removed acoustical material was installed using fasteners or adhesives.

Metal Framing and Furring.   While damaged major metal framing members such as joists, trusses, beams, and girders may be responsible for acoustical ceiling failure, repairing them is beyond the scope of this book. They should be investigated as a possible source of the failure, as discussed earlier, and repaired as necessary. The following paragraphs assume that damaged concrete or metal structure, and solid or supported substrates supporting metal furring has been repaired and presents a satisfactory support system for the furring being repaired.

Most failed metal furring members are removed, and new material installed, because (even when doing so is possible) it is not often reasonable to repair damaged metal furring members. So, most of the references in the following paragraphs to repairing furring mean removing the damaged pieces and installing new pieces.

Materials.   Materials used to repair existing metal furring should match those in place as nearly as possible but should not be lesser in quality, size, or type than those recommended by recognized authorities or required by the building code.

Standards.   In general, repairs should be made in accordance with the recommendations of recognized standards.

Preparation.   Where existing furring members are damaged, the acoustical treatment must be removed to the extent necessary to permit the repairs to be made. Then the damaged existing furring can be removed, along with related suspension systems, if any. Damaged elements that should be removed include members that are twisted, warped, out of alignment, or otherwise unsuitable for use. Existing hangers that are sound, adequate, and suitable for reuse may be left in place, or removed, cleaned, and reused.

Hangers and hanger attachments that have been left in place where suspended ceiling furring has been removed should be examined to verify their adequacy and suitability to support the new furring. Those found to be unsuitable should be removed.

Existing hangers and hanger attachments in good condition may be used to support new ceiling furring. New hangers and hanger attachments should be provided, of course, where existing hangers and attachments have been removed, where existing hangers or attachments are improperly placed or otherwise inappropriate, and where there are no existing hangers.

New hangers should be attached to the structure, never to pipes, conduits, ducts, or mechanical or electrical devices. Existing hangers so attached should not be used.

Misaligned, Warped, or Twisted Furring. When so misaligned, twisted, or warped that it causes damage to an applied acoustical ceiling, metal furring is usually removed and new furring or another support system provided. Where misalignment alone is the problem, however, reinstallation will often offer a solution.

Incorrectly Constructed Framing or Furring. Where the framing or furring was originally built in such a way that the acoustical ceiling becomes damaged, the necessary corrective measures depend on the type of error. The possibilities are many, and the solutions even more numerous.

Each case must be examined to determine the true cause before steps are taken to correct supposed errors. The extent of the damage must be also taken into account. If it might be possible to prevent the failure from recurring by simply refastening the acoustical ceiling, for example, that should be done before expensive reconstruction of the furring is undertaken.

Failed Ceiling Furring. When an acoustical ceiling that is directly applied to metal furring sags or bows, the cause is often due to failure of the furring or its suspension systems.

Metal ceiling support systems fail for several reasons, including loads placed on the ceiling furring from above, such as storage, structure deflection, differential movement in the structure, improper attachment of the support system, or other trauma. Failure includes broken wire hangers and failed hanger attachments.

When an acoustical ceiling failure is due to a suspended furring system failure, repairs may be made by lifting the ceiling back into place and repairing the failed support system. The first step is to stabilize the ceiling. To do so, access to the space above the ceiling must be obtained. When access is not otherwise possible, it will be necessary to remove portions of the acoustical material.

After obtaining access to the space above the ceiling, the next step is to lift the ceiling from below into proper alignment. When the ceiling is once again in its proper position, new hangers or hanger attachments, as necessary, can be installed to support the furring.

When an acoustical ceiling failure is due to directly-attached metal furring that has not been properly attached, or where the furring attachment has failed, repairs may be made by lifting the ceiling into its proper position and refastening the furring to its supports. Of course, if the acoustical material is not removed while the refastening is being done, the new fasteners will probably be visible. If the furring is supported on metal, driving new fasteners through the ceiling will probably not be possible. In the latter two cases, removal of the acoustical material will probably be necessary. Reinstallation may be impracticable, however, if the removed acoustical material was installed using fasteners.

Rusted Furring.   Members that are extensively rusted are probably not worth saving. Minor rust should be removed and the surface recoated with the material originally used. Painted members should be repainted with a compatible paint. Galvanized members should be repaired using galvanizing repair paint.

Metal Suspension Systems.   It might not be feasible to repair failed metal suspension system components. Even when such is possible, for example, it is seldom practicable to clean and repaint severely rusted members. So, much of the time when this text refers to repairing metal suspension systems it means removing the damaged pieces and installing new pieces in their place.

The following paragraphs assume that structure and solid substrate damage has been repaired and that there is satisfactory support for the suspension system being repaired.

Materials.   Components used to repair existing metal suspension systems should match those in place as nearly as possible, but should not be lesser in quality, size, or type than those recommended by recognized authorities or required by the building code. Where interlocking of members is necessary, new members must be made in the same configuration as those removed. Where a metal ceiling installation requires that the grid member be designed in a particular configuration to accept a snap-in or other special installation method, new suspension systems members must be of the same type as those used in the remainder of the ceiling. This usually dictates that the same manufacturer's product be used. Errors in the original application should not be duplicated, however, purely for the sake of matching the original.

Damaged existing materials should not be used in making repairs or in extending the existing surfaces unless they can be satisfactorily repaired.

Hangers and hanger attachments left in place after suspension system grid members have been removed should be examined carefully and their adequacy and suitability verified. Those found to be unsuitable should be removed.

When installing new suspended ceilings grid components, existing hangers and hanger attachments may be used where they are suitable. New hangers and hanger attachments must be provided where existing hangers and attachments are removed, where existing hangers or attachments are improperly placed or otherwise inappropriate, and where there are no existing hangers. New hangers should be attached to concrete, structural steel, or wood framing members. They should never be attached to steel deck, pipes, conduits, ducts, or mechanical or electrical devices, and existing hangers that are so attached should not be used.

New hangers should be plumb in relation to primary suspension members and not in contact with insulation covering ducts and pipes. Hangers should be splayed only where obstructions or conditions prevent plumb, vertical installation. The horizontal forces generated by splaying of hangers should be offset by counter-splaying, bracing, or another suitable method.

When such existing hangers have been removed, and when a lack of such hangers contributed to the failure, additional hangers must be installed to support the grid systems at lights, grilles, registers, and other items, to prevent excessive deflection. In grid systems with a "light" structural classification, lights, grilles, registers, and other items must be supported separately from the ceiling grid.

Installed suspension systems should achieve a tight and level ceiling at the spacings and directions recommended by the manufacturer, and to comply with UL label requirements.

Standards. In general, repairs should be made in accordance with the recommendations of recognized standards, such as those mentioned earlier in this chapter or in "Where to Get More Information" at the end of this chapter, or both, and the standards referenced in the Bibliography as applicable to this chapter.

Preparation. Where existing suspension system components are damaged, the supported acoustical materials must be removed to the extent necessary to permit the repairs to be made. Then the damaged existing suspension system components can be removed, along with their hangers and hanger attachments. Removed damaged elements that are not repairable should be discarded. Damaged elements that should be removed include, but are not limited to, suspension system components that are broken, bent, or otherwise physically damaged, rusted beyond repair, or otherwise unsuitable for reuse; and hangers and hanger attachments that are rusted, broken, or otherwise damaged. Existing hangers, hanger attachments, and grid system components that are sound, adequate, and suitable for reuse may be left in place, or removed, cleaned, and reused.

Hangers, hanger attachments, and grid members that have been left in place where some grid system components have been removed should be examined to verify their adequacy and suitability for use in the repaired suspension system. Those found to be unsuitable should be removed.

Existing hangers and hanger attachments in good condition may be used to support new grid system components. New hangers and attachments should be installed, of course, where existing hangers and attachments have been removed or are improperly placed or otherwise inappropriate, and where there are no existing hangers. New hangers should be attached to the structure, never to pipes, conduits, ducts, mechanical or electrical devices, or steel deck. Existing hangers so attached should not be used.

Misaligned, Bent, or Twisted Suspension System Components. Concealed metal suspension system components that are misaligned, bent, or twisted enough to cause damage to other components or the acoustical materials can sometimes be left in place and the other components reattached in such manner that the poor condition of the damaged suspension system components is overcome. When repair to the exposed components or acoustical materials alone will not prevent failure from recurring, it will be necessary to remove the damaged components and provide new materials. Sometimes, removed metal suspension system components that were only misaligned can be properly reinstalled.

Incorrectly Constructed Suspension System Components. When an existing suspension system was built in such a way that the acoustical materials become damaged, the necessary corrective measures depend on the type of error. The possibilities are many, and the solutions even more numerous.

Each case must be examined to determine the true cause before steps are taken to correct supposed errors. The extent of the damage must be also taken into account. If it is possible to prevent the failure from recurring by simply rehanging the suspension system components, for example, that should be done before expensive reconstruction of the suspension system is done.

Failed Suspension System Components. Metal ceiling suspension systems may fail for several reasons, including loads placed on the suspension system components from above, such as those from items being stored there, structure deflection, differential movement in the structure, improper attachment of the support system, or other trauma.

When an acoustical ceiling failure is due to suspension system component failure, repairs must include removing the acoustical ceiling materials and repairing the failed suspension system.

Damaged Finishes. Scratched and abraded painted surfaces can be repainted using any paint type recommended for use on metal surfaces. Preparation should be done as recommended by the paint manufacturer. Refer to Chapter 5 for a discussion of paints, including materials, surface preparation, and materials use. The paint used must be compatible with the existing finish. Should there be any doubt about the compatibility of materials, it is best to contact the ceiling grid manufacturer for advice. Except for minor touch-up of scratches, it is usually better to refinish an entire member rather than just paint the damaged area, so that repairs will be less apparent.

Clear coatings on natural aluminum finishes can usually be repaired using the same clear coating originally used, but such repairs may be visible. Such clear coatings are usually, but not always, lacquer.

Repair of anodized aluminum, plastic coatings, and fluoropolymer finishes should be made in accordance with the grid manufacturer's and finisher's recommendations. Similar coatings may not respond alike to the same refinishing method. Techniques that work well with one finish may be harmful to another, similiar finish.

Damaged finishes on metal members that are not exposed may be repaired using the same type of material that was used in the original finish. Painted finishes may be repainted. Galvanized surfaces may be recoated using galvanizing repair paint.

Cleaning.   Exposed ceiling grid components can usually be cleaned using a soft brush or vacuum cleaner. Marks may be removed with an art gum eraser or a wall cleaner. Care must be exercised to not damage adjacent acoustical materials while cleaning metal grid system components. It is best to use as little water and cleaning fluids as possible. Marks that cannot be removed may be covered with a small amount of paint, as discussed in the preceding paragraphs.

Acoustical Materials

Materials.   Acoustical materials for use in patching and extending existing acoustical ceilings should exactly match the existing materials. Care must be exercised when selecting materials from a different manufacturer than the one that made the materials originally used. Products of the same type, form, pattern, grade, color, and light reflectance may vary significantly in appearance.

Cleaning.   The manufacturer's instructions should be obtained and followed when cleaning acoustical ceiling materials. The following suggestions are general in nature and should not take precedence over the manufacturer's instructions.

Dust and loose dirt may be removed from most acoustical materials using a soft brush or a vacuum cleaner with the type of attachments used for cleaning upholstery. Brushing should be done with a light touch and all in the same direction. Hard brushing will drive dirt and dust into the perforations and, in softer materials, into the acoustical material itself, making it harder to remove.

Marks that are not removed by light dusting, such as pencil marks and smudges (Fig. 3-15), may be cleaned off most painted acoustical materials using an art gum eraser or a wallpaper cleaning product. The wallpaper cleaner should be unused.

On most painted materials, large areas of soiling may be washed using a mild soap (not detergent) and a damp cloth or sponge. Acoustical materials must not be immersed or soaked. Soap used in cleaning must be removed with a damp cloth or sponge containing clear water. Some types of acoustical material finishes may withstand more scrubbing than others without damage, but each should be tested in a small, inconspicuous area first. It is not always prudent to accept blindly the claims of a manufacturer.

Surfaces with a natural finish offer different cleaning problems than painted surfaces. They should be cleaned only according to the manufacturer's instructions.

Figure 3-15 It may be possible to remove smudges such as these by using vacuum cleaner and brush, but it will probably require harsher methods. (*Photo by author.*)

Surface Damage. Minor scratches, nicks, and chips in painted composition acoustical materials may be concealed using chalk, pastel, typist correction fluid, shoe polish, or a small amount of paint. Similar materials will also conceal minor scratches in metal ceiling materials that have a painted finish.

Most painted acoustical materials can be repainted without significant loss of acoustical proficiency. The manufacturer should be consulted, however, before proceeding. Most acoustical materials can be repainted using vinyl latex paint, but the acoustical material manufacturer may have specific recommendations regarding which material to use. The gloss of the paint should, of course, match the gloss of the original finish on the acoustical material.

Paint may be applied to most acoustical materials (Fig. 3-16) using either spray, roller, or brush. Some patterns may significantly lose acoustical performance, however, when spray painted. When there is a doubt, it is always best to contact the acoustical material manufacturer for advice before selecting the method to use. Spray or roller application take the least time, of course.

Figure 3-16 Paint will sometimes salvage stained acoustical ceiling materials such as this panel, if the panel can be dried. (*Photo by author.*)

Regardless of the paint material and application method used, the applicator must be careful to ensure that the openings in the acoustical material are not clogged with paint. Blocking the perforations will cause a significant loss of acoustical performance.

Acoustical materials to be painted should be removed from the ceiling, thoroughly cleaned of dust, dirt, and other foreign particles, and permitted to dry completely. One coat is normally sufficient if the selected paint has good hiding qualities, but two coats may be necessary in some cases, especially if it is necessary to thin the paint to prevent sealing the perforations in the acoustical material. Painted acoustical materials should be permitted to dry on a flat surface before reinstallation.

Clear coatings on natural aluminum finishes can usually be repaired using the same clear coating originally used, but such repairs may be visible. Such clear coatings are usually, but not always, lacquer.

Repair of anodized aluminum, plastic coatings, and fluoropolymer finishes should be made in accordance with the grid manufacturer's and finisher's recommendations. Similar coatings may not respond alike to the same refinishing method. Techniques that work well with one finish may be harmful to another, similar finish.

Acoustical Materials Installation. Removed units should generally be reinstalled in the same location from which they were removed, so that the direction of the pattern and the appearance of the ceiling is not altered from that of the original application. When the ceiling installation has been completed, the entire ceiling area should be covered with acoustical material. Where necessary, acoustical materials should be cut around fixtures, penetrating hangers, and accessories. New border units should be cut and joints should be made neatly to fit snugly into the grid, leaving no voids in the finished ceiling.

Installing New Acoustical Ceilings over Existing Materials

General Requirements. The following paragraphs contain some suggestions for installing new acoustical ceilings, including their furring and suspension systems, in existing spaces. Because the suggestions are meant to apply to many situations, they might not apply to a specific case. In addition, there are many possible cases that are not specifically covered here. When a condition arises in the field that is not addressed here, advice should be sought from the additional data sources mentioned in this book. Often, consultation with the manufacturer of the materials being installed will help. Sometimes, it is necessary to obtain professional help (see Chapter 1).

Under no circumstances should these recommendations be followed without careful investigation and the application of professional expertise and judgment.

The discussion earlier in this chapter about new acoustical ceiling materials and installations also apply to new ceilings installed in existing spaces. There are, however, a few additional considerations to address when the new ceiling is installed in an existing space. The following paragraphs address those concerns. Some, but not all, of the requirements discussed earlier are repeated here for the sake of clarity. It is suggested that the reader refer to earlier paragraphs under this same heading ("Acoustical Ceilings") for additional data. Specific applicable headings include:

- Acoustical Materials: "Acoustical Materials."
- Furring and Metal Suspension Systems: "Support Systems."
- Accessories and Miscellaneous Materials: "Accessories and Miscellaneous Materials."
- Fire Ratings: "Fire Performance."
- Furring, Suspension System, and Acoustical Material Installation: "Installation."

The existing areas where new acoustical ceilings will be installed should be inspected to verify that existing ceilings that are to be removed have been removed, work to be concealed by the ceiling has been completed, the area is ready to receive the ceilings, and the actual conditions are the same as those expected.

Hangers and hanger attachments left in place when existing ceilings were removed should be inspected to verify their adequacy and suitability for support of new suspension systems. Those found to be unsuitable should be removed and new hangers and attachments installed in their place.

Acceptable existing hangers and their attachments may be used to support new ceilings in existing spaces. New hangers and hanger attachments must be installed, of course, when existing hangers and attachments have been removed, where existing hangers or attachments are improperly placed or are otherwise inappropriate, and where there are no existing hangers. Removed hangers and hanger attachments should not generally be reused, unless their condition is exceptional. Such removed items may have hidden damage. Hangers should be attached to structural elements, not to steel deck, pipes, conduits, ducts, or mechanical or electrical devices. Existing hangers so attached should not be reused.

Installing New Wood Furring over Existing Materials. Where a new acoustical ceiling is to be applied in an existing wood-framed space, a new wood furring system is sometimes used. Installing new, and repairing and extending existing, structural framing systems are beyond the scope of this book.

Even in new buildings, wood furring is always applied over something else. The problems vary only slightly when the something else is old. More shimming may be necessary when new furring is applied on existing substrates, of course, and sometimes an existing surface will be in such poor shape that it cannot be satisfactorily furred. When that happens, it may be necessary to remove the existing construction completely and build it new.

It is not always necessary to remove existing trim when installing a new furring system. Items that will be concealed in the finished installation can be left in place.

New wood furring can be applied directly onto ceilings that are themselves directly applied to wood framing, such as joists or trusses, when the framing is sufficiently strong to hold both new and existing ceilings. New furring strips should be laid perpendicular to the existing framing, at a spacing appropriate for the new acoustical ceiling, and nailed or screwed through the existing finish to the existing framing. The nails or screws should extend into the existing framing at least 1-1/4 inches.

New wood ceiling furring can also be suspended from existing wood framing. The new furring should be 2 by 4's placed at the appropriate spacing for the new acoustical material. Where the existing framing is exposed, the new furring can be hung by wires from fasteners driven into the framing. The wire may be tied to nails driven downward into the existing framing, but eye screws are preferable.

Where the existing framing or furring is both sound and strong enough, it is sometimes preferable to leave an existing ceiling in place and hang the new ceiling through it from the existing framing or furring. This may be done in two ways. Large eye-screws can be driven through the existing ceiling into the left-in-place wood framing or furring, and the new suspended ceiling furring members hung from them. Such eye-screws should penetrate at least a full inch into the existing framing or furring. The second method is to punch sufficient holes of large enough size in the existing ceiling to permit suspending the new system from the existing structural framing.

In every case, care must be taken to ensure that new hangers are atached to sound framing elements that are of sufficient strength. Hangers should not be attached to pipes, conduits, ducts, or mechanical or electrical devices, or to framing or furring that is damaged or too weak to carry the loads to be supported.

Installing New Metal Furring over Existing Materials. Where a new acoustical ceiling is to be applied in an existing space, a new metal furring system is sometimes used. Most of the time, however, it is better in such cases to install a new suspended furring system covered with a backing board and fasten the new acoustical material directly to it using fasteners or adhesives. Installing a suspended gypsum backing board is beyond the scope of this book, but is discussed in the book in this series entitled *Repairing and Extending Finishes, Part I.*

Even in new buildings, metal furring is always applied over something else. The problems vary only slightly when the something else is old. More shimming may be necessary when new furring is applied on existing substrates, of course, and sometimes an existing surface will be in such poor shape that it cannot be satisfactorily furred. When that happens, it may be necessary to remove the existing construction completely and build it new.

New metal furring can sometimes be applied directly onto ceilings that are themselves directly applied to wood or metal framing or furring, when the existing framing or furring is sufficiently strong to hold both new and existing ceilings. New furring members should be laid perpendicular to the existing framing or furring, at a spacing appropriate for the new acoustical ceiling, and fastened through the existing finish to the existing framing or furring.

It may not be practicable, however, to install new metal furring over an existing ceiling supported directly on metal furring or framing. There is usually nothing readily available to which the new furring can be fastened. Fastening to the existing ceiling or metal furring is not appropriate. It will probably be necessary in such cases to remove at least enough of the existing ceiling so that the new furring can be fastened to the sructure. It may be possible to remove the existing ceiling and reuse the existing furring, however, if the furring is sufficiently stable, strong, and at the correct spacings for the new ceiling. At any rate, the latter possibility should be explored before an entirely new furring system is installed.

New metal ceiling furring can be suspended from existing wood or metal framing, but a better solution is the method mentioned earlier involving a backing board. When a ceiling exists, it may be left in place and only sufficient material removed to permit fastening the new support system to the structure.

In every case, care must be taken to ensure that new hangers are attached to sound framing elements that are of sufficient strength. Hangers should not be attached to pipes, conduits, ducts, or mechanical or electrical devices, or to framing that is damaged or too weak to carry the loads to be supported.

Installing New Metal Suspension Systems over Existing Materials.   The principles discussed in this book for installing a new suspension system in a new building are applicable when installing a new suspension system in an existing building.

Even in new buildings, suspension systems are always hung from something else. The problems vary only slightly when the something else is old.

It is not always necessary to completely remove existing ceiling materials when installing a new suspension system over an existing surface. Access to the supporting structure must be gained, of course, but items that do not block that access and will be concealed in the finished installation can be left in place, unless a code or other restriction requires their removal.

Some new ceiling suspension systems can be hung directly from the existing framing. Others require the imposition of carrying channels. Even systems that are usually hung directly from the structure may need carrying channels when hangers cannot be properly spaced for direct suspension. The requirements for carrying channel, furring member, and hanger spacing are the same as was discussed earlier in this chapter for new suspension systems. When the existing framing is exposed, the new furring should be installed exactly as it would be in all-new construction.

It is sometimes preferable to leave an existing ceiling in place, penetrating it only as necessary to install hangers. The number of hangers, and thus penetrations through the existing ceiling, can sometimes be reduced by using heavier carrying channels and hangers, and thus increasing the spacing of the hangers. The suspension system manufacturer and the referenced ASTM standards should be consulted before such special applications are attempted, however.

In every case, care must be taken to ensure that new hangers are attached to structural framing elements. Hangers should not be attached to steel decks, pipes, conduit, ducts, or mechanical or electrical devices.

Installing Tile and Panels.   Requirements stated earlier for all new ceilings in new spaces apply as well to ceilings in existing spaces.

Integrated Ceilings

Integrated ceilings consist of acoustical materials, a suspension system, outlets (grills, vents, or vanes) for air distribution, and electrical devices, such as lights, combined into a single system the components of which are compatible visually and mechanically.

System Components

Acoustical Materials.   The acoustical materials used in integrated ceilings include the composition and metal acoustical materials listed in Federal Specification SS-S-118 and discussed earlier in this chapter under ''Acoustical

Ceilings." All of the statements made there apply to acoustical materials used in integrated ceilings.

Besides the types of acoustical materials specifically listed in Federal Specification SS-S-118, the other types of materials mentioned earlier that may be used in acoustical ceilings may also be used in integrated ceilings. Included are ceramic-faced units and units made from asbestos, plastic, gypsum, and wood fiber. Integrated ceilings may also have wood, plastic, or sheet metal panels or tiles, but many of them cannot properly be classified as acoustical materials.

Support Systems. Most statements made earlier in this chapter about suspension systems for acoustical ceilings apply as well to the suspension systems used in integrated ceilings. One major difference is that suspension systems for integrated ceilings are usually designed to support all of the components of the ceiling, including such devices as light fixtures. Another major difference is that the grid members of an exposed-grid integrated system are often designed to provide air supply or return.

Grids for integrated ceilings may be either concealed or exposed. In linear metal integrated ceilings, the supporting members usually are not exposed. Grids are also usually concealed in integrated ceilings where the acoustical or decorative panels, tiles, pans, or strips snap into the system.

Air Diffusion. A major difference between acoustical ceilings and integrated ceilings is that the latter are designed to contain air supply and, often, air return outlets of the heating, ventilating, and air conditioning system. The effect of including such air distribution elements in the ceiling system can result in very effective concealment of diffusers and grills that are otherwise often ugly and obtrusive. Some integrated systems employ air supply and return slots that cannot be discerned as such from floor level.

All integrated ceilings employ boots or manifolds to connect the integral air supply and return elements to the building's ductwork. Connections from the ceiling to the ductwork are usually flexible.

Lighting. Spaces with acoustical ceilings are often lit by standard fixtures not specifically designed for the purpose, which are mounted on the surface of the ceiling or are recessed into the ceiling. Often recessed fixtures in acoustical ceilings are standard units laid into the suspension grid. Some integrated ceilings also use standard recessed or surface-mounted fixtures. But one of the more attractive features of many integrated ceilings is that they include light fixtures that have been designed specifically for the particular integrated ceiling system. The result is often a more attractive installation than results from many acoustical ceilings.

Lighting design is usually an integral part of the design of an integrated ceiling. Fixture locations and types are sometimes limited, but many integrated ceiling designs result in effective and comfortable lighting. Care must be exercised in some cases, especially where standard fixtures are used, to prevent glare.

Acoustical Performance.   The statements made earlier in this chapter about the acoustical performance of acoustical ceilings apply as well to integrated ceilings. Integrated ceilings are often used in open-plan spaces where the design should include acoustical materials having high NRC values. Such installations are sometimes accompanied by so called "white noise" generators to mask background noises.

Fire Performance.   The statements made earlier in this chapter about the fire performance of acoustical ceilings apply as well to integrated ceilings. There are some additional requirements associated with integrated ceilings, however. The fire resistance classification of the system must take into account the electrical and mechanical components of the system. When an integrated ceiling must be time-rated for fire resistance, the limitations may be severe. Light fixture and mechanical opening sizes and locations may be restricted. The method of enclosing such devices will probably be dictated by the rating designation. The appearance of the installed system may be affected drastically by such limitations. A fire-rated integrated ceiling may not look much like a nonrated system using the same basic components.

Installation

Unlike some acoustical ceilings, integrated ceilings are seldom fastened directly to wood or metal framing or furring. Most integrated ceiling installations, almost by definition, include some sort of metal suspension system.

Most of the requirements suggested earlier in this chapter for installing suspension systems for acoustical ceilings apply as well to integrated ceilings. Suspension systems, for example, may be either direct-hung or indirect-hung, and the same ASTM standards apply. Grids may be either concealed or exposed, depending on the particular ceiling system. In every case, the installation should follow the integrated ceiling manufacturer's recommendations.

Light fixtures and air distribution components should be correctly installed so that their functions are properly carried out. Their installation must be coordinated with the mechanical and electrical system installations.

Why Integrated Ceilings Fail

Integrated ceilings may fail for the same reasons acoustical ceilings fail. Refer to the heading "Why Acoustical Ceilings Fail" earlier in this chapter for a detailed discussion. The numbered lists that appear there under various subheadings apply also to failures in integrated ceilings. Possible failure causes include those discussed in Chapter 2, such as structure failure, structure movement, solid substrate problems, supported substrate problems, and other building element problems.

Possible failure causes also include those listed under "Bad Materials," "Improper Design," "Bad Workmanship," "Failing to Protect the Installation," and "Poor Maintenance Procedures."

In addition to the causes listed under "Acoustical Ceilings," integrated ceilings may experience failure due to the following:

Improper Air Distribution Design or Installation. Reasons for failure include the following:

1. Failing to properly design the air distribution system.
2. Failing to follow the manufacturer's instructions.
3. Failing to install the air distribution elements in accordance with the design.
4. Failing to properly install the air distribution system.

Improper Lighting Design or Installation. Reasons for failure include the following:

1. Failing to properly design the lighting system.
2. Selecting the wrong light fixtures.
3. Failing to follow the manufacturer's instructions.
4. Failing to install the lighting system elements in accordance with the design.
5. Failing to properly install the lighting system.

Evidence of Failure

The paragraphs under this heading follow the same principles as those under "Evidence of Failure" under the main heading "Acoustical Ceilings." In addition to the failures listed there, integrated ceilings are also subject to the types of failure listed below. The causes listed below are discussed in the previous paragraphs under "Why Integrated Ceilings Fail" and the numbered causes here are those listed under that heading.

Poor Heating, Ventilating, or Air Conditioning Performance.   Poor performance may include hot or cold spots, uneven air distribution, drafts, air noises, or temperatures that are too high or low. Any of the following may be at fault.

- Improper Air Distribution Design or Installation: 1, 2, 3, 4.

Poor Lighting Performance.   Poor performance may include glare, low light levels, dark spots, or uneven distribution. Any of the following may be at fault.

- Improper Lighting Design or Installation: 1, 2, 3, 4, 5.

Cleaning, Repairing, Refinishing, and Extending Integrated Ceilings

Except where specifically not applicable (wood furring requirements, for example), the discussion under "Cleaning, Repairing, Refinishing, and Extending Acoustical Ceilings" earlier in this chapter applies to integrated ceilings as well. To maintain the same appearance and function, of course, it is necessary to use the same brand and specific part as was used in the original installation when replacement of integrated ceiling parts is necessary.

Air diffusion element and lighting fixture cleaning, repairing, and refinishing should be done in accordance with the manufacturer's instructions.

Installing New Integrated Ceilings over Existing Materials

The discussion in "Installing New Acoustical Ceilings over Existing Materials" earlier in this chapter applies to integrated ceilings as well. It is even more true for integrated ceilings than it is for acoustical ceilings that, while general installation principles should be in accordance with accepted ASTM and other industry standards, the design of the particular integrated ceiling system dictates specific installation methods.

The manufacturer's instructions should be followed carefully when installing a new integrated ceiling in an existing space.

Acoustical Wall Panels and Baffles

Acoustical wall panels include standard and custom spline-mounted and back-mounted panels. Baffles are either double- or single-thickness. Wall panels and baffles are faced with either standard, custom, or owner-furnished facing materials.

Demountable partitions and movable partitions, which may also be finished with acoustical panels, are beyond the scope of this book. When an acoustical material associated with one of them fails, however, the problem and its solution may be very close to those discussed in this chapter. Therefore, the discussion here may offer a preliminary suggestion as to the nature of the problem and its solution. However, since all such partitions are proprietary, it is usually best to contact the manufacturer of the partition that has failed for advice before attempting to make repairs.

Also beyond the scope of this book are the several available acoustical wall systems in which a fabric is stretched over an applied layer of acoustical material. In them, the fabric is either held by proprietary plastic accessories or fastened to wood trim members. When a failure is discovered in such a system, the problem and its solution may closely resemble those discussed in this chapter. But, before any attempt is made to correct a failure in one of those systems, it is best to first contact the manufacturer of the system for advice.

Panel and Baffle Materials and Fabrication

Most acoustical wall panels and baffles consist of a fabric or vinyl facing over mineral, glass, or wood fiberboard. Materials and fabrication may vary, however, between spline-mounted and back-mounted panels.

Metal acoustical wall panels and baffles are also available.

Spline-mounted Panels.   The core of most spline-mounted panels is faced on one side only because the panels are mounted with butted joints. The finish is wrapped around the long edges back to the line of the kerf, however. The short panel ends are unfinished.

Panel Cores.   The cores of spline-mounted panels can be either perforated water-felted mineral fiberboard with a fairly high density, such as 17 to 20 pounds per cubic foot; low-density mineral fiberboard; glass fiberboard with a nominal density between 6 and 8 pounds per cubic foot; or high-density wood fiberboards. The long edges of spline-mounted panels are kerfed and rabbeted to receive mounting splines (Fig. 3-17).

Accessories.   Mounting and trim accessories for spline-mounted panels include H- and J-shaped splines, and J-, H-, and angle-shaped moldings, as shown in Figure 3-18. Some manufacturer's accessories may vary from those shown. Accessories may be either extruded aluminum or plastic.

Wood battens are sometimes used to cover the joints, and sometimes to support panels. Wood battens are usually decorative hardwood, but may be painted softwood in some installations.

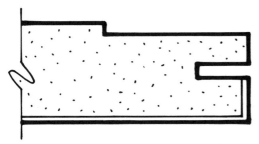

Figure 3-17 Typical long edge of a spline-mounted acoustical wall panel.

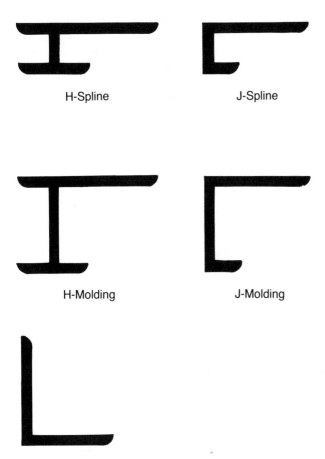

H-Spline

J-Spline

H-Molding

J-Molding

Angle Molding

Figure 3-18 Some typical accessories for spline-mounted acoustical wall panels.

Panel Size and Thickness.   Spline-mounted panels are either 9 or 10 feet long and either 24, 30, or 48 inches wide. High-density fiberboard panels vary in thickness from 1/2 inch to 1-1/8 inches. Glass fiberboard and low-density fiberboard panels vary in nominal thickness from 1 inch to 1-1/8 inches.

Back-mounted Panels.   The core face that will be exposed after the panel is installed and the edges of back-mounted panels are usually finished.

Panel Cores.   The core of back-mounted panels may be glass fiberboard with a nominal density between 4 and 7 pounds per cubic foot or high-density wood fiberboards. Glass fiber panels may have impact-resistant faces which usually consist of a 1/8-inch-thick layer of 18 pound per cubic foot density glass fiberboard laminate to the face of the core board.

The edges of the cores of back-mounted panels are either chemically hardened to reinforce the edge and help prevent warpage and damage, or framed with aluminum, zinc-coated steel, or wood.

Panel edges that are not framed are usually wrapped with the facing material. Edge shapes of panels with wrapped edges may be square, beveled, radiused, notched, mitered, or a combination of those (Fig. 3-19).

Panel corners may be either square or rounded.

Some panels are faced on the back-side with a thin aluminum foil to prevent air passage.

Accessories.   Mounted accessories for back-mounted panels include conventional Z-clips, strip-clips which consist of Z-clips mounted on a continuous strip, and magnetic pads (Fig. 3-20). Other proprietary clip assemblies are also used. Panels are also mounted using hook and loop panels (Velcro), and sometimes using adhesives.

High-density wood fiber panels are sometimes installed using screws through the panels.

Wood battens are sometimes used to cover the joints between panels, and sometimes to hold the panels in place.

Panel Size and Thickness.   Back-mounted panels vary in thickness from 1/2 inch to 3 inches. Panels are available in a large variety of sizes from 15 inches square up to 5 by 12 feet.

Metal Wall Panels.   Some of the metal ceiling products discussed earlier in this chapter under ''Acoustical Ceilings'' are sometimes used on walls, but here we will discuss only products that are specifically designed for use on walls. Such panels are available in both aluminum and galvanized steel.

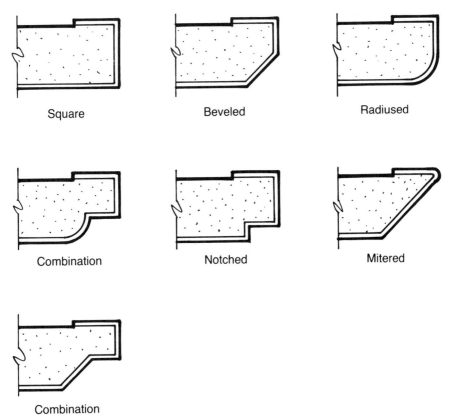

Square   Beveled   Radiused

Combination   Notched   Mitered

Combination

Figure 3-19 Some typical edge profiles for back-mounted acoustical wall panels.

Metal wall panels are available flat or shaped (corrugated, for example), perforated or not perforated, and either plain, embossed, or textured. Most panels come with glass fiber or mineral wool sound-absorbing pads. Some pads are wrapped or faced with materials such as polyvinyl chloride. Others are bare.

Special brackets and other accessories are required for some metal acoustical wall panel installation.

Generally, available metal wall panel sizes vary from 18 inches to 4 feet wide and from 18 inches to 12 feet long. Some may be smaller or larger.

Baffles. A baffle is often nothing more than a standard acoustical wall panel of the back-mounted type that is finished on both faces and all edges. Some baffles are the equivalent of two acoustical wall panels laminated

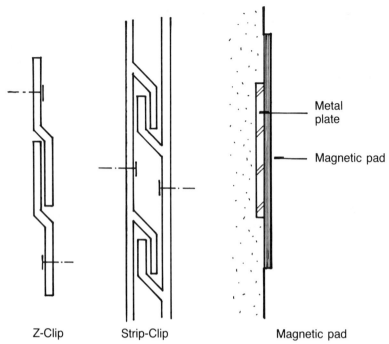

Z-Clip   Strip-Clip   Magnetic pad

Figure 3-20 Typical mounting accessories for back-mounted acoustical wall panels.

back to back. Some double-thickness baffles are completely encased with the finish material. Others have a joint between the two sides of the baffle. Most baffles are flat panels, but cylindrical, octagonal, triangular, and other-shaped baffles are also available. Their basic panel construction is similar to that of flat baffles.

Baffle edges may be finished in the same fabric or vinyl as the face, or framed in wood or metal. Edge configurations are similar to those for back-mounted acoustical wall panels.

Metal baffles may be flat panels similar to the metal acoustical ceiling materials discussed under "Acoustical Ceilings," or more like the metal wall panels discussed earlier under "Metal Wall Panels."

Baffles are made to be mounted or suspended by special hangers either built into or attached to the baffle. Some baffle hangers are nothing more than eye bolts mounted in the narrow edge of the panel or grommets through the panel face. Others use more elaborate hanger systems that vary from manufacturer to manufacturer. Hangers may be cable, wire, decorative chain, or almost invisible nylon fishing line. Sometimes hangers are solid metal bars. Some baffles are designed to be clipped to overhead framing.

Wall Panel and Baffle Finishes

Fabric. Standard panels and baffles are available with both woven and nonwoven synthetic fabrics, the majority of which are largely or entirely polyester. Other fabrics are also used. Some manufacturers state that almost any upholstery-weight fabric may be used, but fire-resistance requirements may not permit the use of just any fabric. Some sources state that some wool, nylon, metallic, rayon, and acetate fabrics are not suitable.

To determine the actual fabric used in an existing installation, it is usually necessary to contact the panel or baffle manufacturer. When the fabric used was furnished by the owner, as is permitted with some panels, it will be necessary to ascertain the name of the fabric used from someone other than the panel manufacturer.

Vinyl. Perforated vinyl is often used as a facing for acoustical wall panels, especially in spaces, such as classrooms, corridors, and the like, where the material is likely to become soiled quickly. It is also sometimes used for facing on baffles. In older installations the vinyl may have had large holes punched in it to provide the proper acoustical performance. Vinyl in newer installations will probably have been punched using a process called microperforating, which eliminates large holes.

Metal. Most metal panels designed for wall panel or baffle use are painted with either acrylic or polyester baked-on paint. Other finishes may be available.

Other Finishes. More utilitarian baffles may be faced or wrapped with polyethylene or glass fiber cloth.

Acoustical Performance. Acoustical wall panels and baffles are generally classified by their Noise Reduction Coefficient (NRC). Panels and baffles fall into two general categories. The first includes panels that will absorb between 60 and 70 percent (NRC 0.60 to 0.70) of the sound that strikes them. They are used in normal-sized rooms, such as conference rooms and offices. The secondary category includes panels that absorb between 80 and 95 percent (NRC 0.80 to 0.95) of the sound that strikes them. They are used in large spaces, such as open-plan offices and spaces where high noise levels are generated, such as computer rooms, swimming pools, and gymnasiums.

Where sound control is an essential element in an existing building, the advice of an acoustical specialist should be sought before changes or repairs are made.

Support Systems

Spline-mounted acoustical wall panels may be directly attached to supported substrates or supported by wood or metal furring applied over wood or metal studs or solid substrates. Back-mounted acoustical wall panels are usually directly attached to supported substrates using one of the accessories discussed later in this chapter. Metal acoustical wall panels may be directly attached to solid or supported substrates, or supported by metal furring applied over studs or solid substrates.

Wood Furring.   Spline-mounted acoustical wall panels and metal acoustical wall panels are sometimes installed over horizontal wood furring (Fig. 3-21). Such wood furring is most often used when the substrates are solid ones, such as concrete or masonry, and the substrates are too irregular to directly apply the splines and accomplish a finished surface within acceptable tolerances. Wood furring may also be used over existing wood framing that is not in proper alignment to receive acoustical wall panels, but such use is rare in new buildings. Wood furring may also be used over wood framing for the support of acoustical baffles when the framing is not at the proper spacings or locations for the baffle hangers.

Back-mounted acoustical wall panels are usually mounted directly onto a supported substrate, such as plaster or gypsum board. Where wood furring is required in such installations, it is used to fur the supported substrate and not the back-mounted acoustical panels.

Wood framing and furring that supports plaster and gypsum board are not addressed in this book in any appreciable detail, but are discussed in detail in the book in this series entitled *Repairing and Extending Finishes, Part I.*

When wood furring is used, it should comply with the building code and the standards and minimum requirements of generally recognized industry standards.

Wood.   Probably the single largest cause of problems with acoustical wall panels applied over wood furring is the wood not being properly cured when installed. Wood for furring should be seasoned lumber with a 19 percent maximum moisture content at the time of dressing. Lumber with a moisture content in excess of 19 percent can be expected to change in size by 1 percent for each 4 percent reduction in moisture content. As the wood changes in size, it will usually also warp and twist, especially when held in place at the ends, as is the case in framing and furring members.

Softwood used for furring should comply with the U.S. Department of Commerce's *PS 20,* and the National Forest Products Association's *National Design Specifications for Wood Construction.* Each piece of lumber should

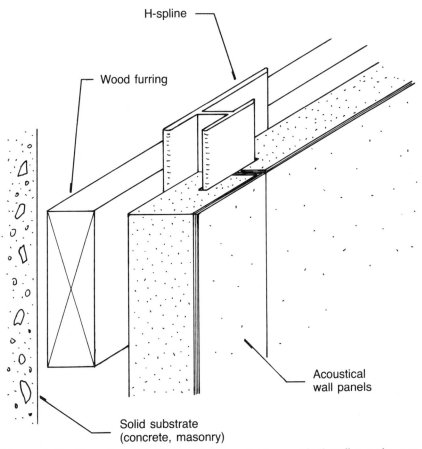

Figure 3-21 Mounting detail for splint-mounted acoustical wall panels over a solid substrate.

be grade-stamped by an agency certified by the Board of Review, American Lumber Standards Committee.

Lumber for furring may be any of a number of available species, including, but not limited to, Douglas fir, Douglas-fir-larch, hem-fir, southern pine, spruce-pine-fir, and redwood.

In newer buildings, actual wood sizes will probably be in accord with the U.S. Department of Commerce's *PS 20,* and the lumber will probably be surfaced on all four sides. In older buildings, though, lumber may have a different actual dimension than is normal today for the same nominal size. A 2 by 4, for example, may actually be 2 inches by 4 inches, instead of 1-1/2 by 3-1/2 inches, which is the standard today. Or, it may be 1-5/8 by 3-5/8 inches, which was the standard for many years. Older furring and framing lumber may not be surfaced on any sides, or may be surfaced only on the sides to which finishes are applied.

Most furring that directly supports acoustical wall panels or to which acoustical baffle hangers are attached will not be pressure preservative treated. Sometimes, furring in contact with masonry or concrete will be pressure preservative treated, however.

Treated lumber should comply with the applicable requirements of the applicable standards of the American Wood Preservers Association (AWPA) and the American Wood Preservers Bureau (AWPB). Each treated item should be marked with an AWPB ''Quality Mark,'' but the markings may not be visible after installation.

Most recently-treated furring will have been treated in compliance with AWPA standard C2, using water-borne preservatives complying with AWPB standard LP-2. Some installations might be treated using other AWPA or AWPB recognized chemicals. Older applications may have been treated in accordance with other standards and using other chemicals, sometimes even creosote.

Most furring that directly supports acoustical wall panels, or from which acoustical baffles are hung, will not be fire-retardant treated. Furring where fire-rated construction is required by the building code may be treated, however.

Fire-retardant treated lumber should comply with AWPA standards for pressure impregnation with fire-retardant chemicals, and should have a flame spread rating of not more than 25 when tested in accordance with Underwriters Laboratories, Incorporated (UL) Test 723 or ASTM Standard E 84. It should show no increase in flame spread and no significant progressive combustion when the test is continued for 20 minutes longer than those standards require. The materials used should not have a deleterious effect on connectors or fasteners.

Treated items that are exposed to high humidities should be treated with materials that show no change in fire-hazard classification when subjected to the standard rain test stipulated in UL Test 790.

Fire-retardant treatment chemicals should not bleed through or adversely affect the acoustical wall panel materials used.

Each piece of fire-retardant treated lumber should have a UL label, but the labels may not be visible after application.

Miscellaneous Materials. The following miscellaneous materials are necessary for wood furring installation:

- Rough hardware, metal fasteners, supports, and anchors.
- Nails, spikes, screws, bolts, clips, anchors, and similar items of sizes and types to rigidly secure furring members in place. These items

should be hot-dip galvanized or plated when in contact with concrete, masonry, or pressure treated wood or plywood, and where subject to high moisture conditions or exposed.

- Suitable rough and finish hardware as necessary.
- Bolts, toggle bolts, sheet metal screws, and other suitable approved anchors and fasteners. These should be located not more than 36 inches on center to firmly secure wood furring in place. Nuts and washers should be included with each bolt. Anchors and fasteners set in concrete or masonry should be hot-dip galvanized, and should be types designed to be embedded in concrete or masonry as applicable, and to form a permanent anchorage.

Metal Furring. Spline-mounted acoustical wall panels and metal acoustical wall panels are sometimes installed over horizontal hat-shaped galvanized steel furring (Fig. 3-22). Such metal furring is most often used when the substrates are solid ones, such as concrete or masonry, and the substrates are too irregular to directly apply the splines and accomplish a finished surface within acceptable tolerances. Metal furring may also be used over existing metal framing, such as studs, that is not in proper alignment to receive acoustical wall panels, but such use is rare in new buildings. Galvanized or painted steel channel furring may also be used over wood or metal framing for the support of acoustical baffles when the framing is not at the proper spacings or locations for the baffle hangers.

Back-mounted acoustical wall panels are usually mounted directly onto a supported substrate, such as plaster or gypsum board. Where metal furring is required in such installations, it is used to fur the supported substrate and not the back-mounted acoustical panels.

Metal framing and furring that supports plaster and gypsum board are not addressed in this book in any appreciable detail, but are discussed in detail in the book in this series entitled *Repairing and Extending Finishes, Part I*.

Metal furring that directly supports acoustical wall panels or from which acoustical baffles are hung should comply with the building code and the standards and minimum requirements of generally recognized industry standards.

Metal furring is usually screwed, clipped, or wired to the supporting structure.

Fire Performance

Acoustical wall panels and baffles should be tested for flame spread and smoke-developed ratings as complete panels containing the same components as they will have when installed. They should rate as Class A according

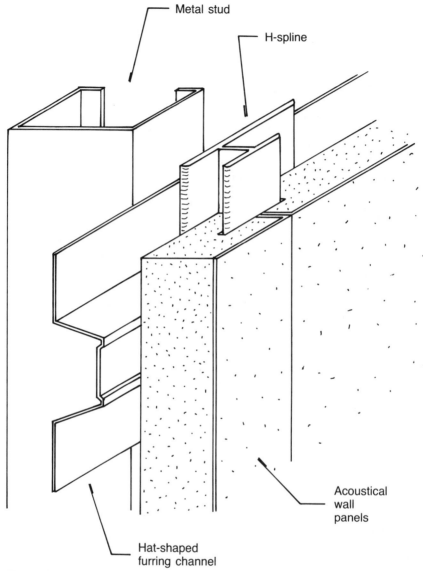

Metal stud

H-spline

Acoustical
wall
panels

Hat-shaped
furring channel

Figure 3-22 Mounting detail for spline-mounted acoustical wall panels over metal studs. Similar details may be used when the wall panels are installed over solid substrates that are irregular.

to ASTM Standard E 84 test, "Surface Burning Characteristics of Building Materials." The flame spread rating of a Class A material ranges from 0 to 25. When the individual components are tested separately, it is difficult to determine a composite value for the entire panel.

Since not all manufacturers have had their acoustical wall panels or baffles tested by independent testing laboratories, such as Underwriters Laboratories (UL), it is often necessary to obtain specific approval of a particular panel or baffle by the local authorities having jurisdiction before using it.

Installation

General Requirements. Acoustical wall panels should be coordinated with electrical outlets and switches, thermostats, and other wall-mounted equipment and devices to ensure that such items are properly accommodated.

Manufacturer's brochures and installation details should be obtained and followed. The manufacturer's recommendations for cleaning and refinishing wall panels and baffles should be obtained and stored in a safe place so that they will be available when maintenance and repair are necessary. They should include precautions against materials and methods which may be detrimental to finishes and acoustical efficiency.

Extra material is occasionally left at the site for the owner to use in making repairs. A search for this material is worthwhile when repairs are needed. Even when there is not enough to make the repairs, the stored material may be in the original cartons, which helps to identify the actual products used in the building.

Acoustical wall panels and baffles should be kept dry before and after use. They should be allowed to acclimate to the temperature and humidity of the space before they are installed to prevent dimensional changes, which might affect the appearance of the installed panels or baffles.

Watermarked, stained, or otherwise damaged units should not be installed.

Acoustical wall panels and baffles should not be installed until the building is weathertight, glazing has been completed, and exterior openings have been closed.

Wet work, such as masonry, concrete, and plaster, should be allowed to dry to the moisture content recommended by the acoustical wall panel or baffle manufacturer, before wall panels or baffles are installed. A temperature of at least 60 degrees Fahrenheit should be maintained in the space during and after installation.

Furring members that abut building structural elements should be isolated from structural movement sufficiently to prevent the transfer of loads into the furring.

Provisions should be made to accommodate movement that occurs across control and expansion joints in the substrates. Furring or panels should not span such joints or be fastened to opposite sides of such joints.

Flat surfaces of furring members and supported substrates should not be warped, bowed, or out of plumb or level by more than 1/8 inch in 12 feet in both directions, or by more than 1/16 inch across any joint or 1/8 inch total along any single member.

Acoustical wall units should be handled carefully to prevent damage to edges and faces.

Hangers for baffles should be properly designed and sufficiently strong to support the baffles. Where splaying of hangers is necessary to avoid ducts, pipes, conduit, or other items, the forces so generated should be compensated for by counter-splaying or another suitable means.

Wood Furring Installation.   Wood furring should be installed as appropriate to support acoustical wall panels and baffles. It should be placed on lines and levels necessary to cause the acoustical panels to fall into the proper location and to properly support baffle hangers at the correct spacings. Where the furring will support directly applied panels or baffles, it should correct unevenness in the supporting structure or substrate. The surface of wood furring to which acoustical panel splines will be fastened should be not less than 1-1/2 inches wide. Generally 2 by 2 lumber is used for furring that is attached to the supporting structure, and 1 by 2 or 1 by 3 lumber is used where furring is laid directly over solid substrates. Nominal 1-inch-thick lumber should not be used over framing because it is too flexible. Other sizes may be found in existing construction, however, especially in older construction. Furring to support baffles should be the proper size to span between the framing members without undue deflection in the furring.

When placed over a solid substrate, wood furring should be installed over a layer of polyethylene film or other vapor-retarding material as an airflow barrier and vapor retarder.

Furring should form a complete system adequate to properly support the acoustical wall panels or baffles. Closers should be installed at edges and openings.

Bolts, lag screws, and other anchors should be used to anchor the furring in place. Generally, fasteners are placed near the ends of furring members and not more than 36 inches on center between. Shorter members, however, should be anchored at 30 inches on center. Bolts should have nuts and washers.

Furring spacings may vary in practice with the substrate and the acoustical panels or baffles being supported. Minimum spacings for all panels are dictated by the panel thickness and size, and may vary somewhat from manufacturer to manufacturer and system to system. Spacing must be such as to adequately support the panels. Spacing of furring for baffle hangers should be such that the baffle hangers fall in the correct locations.

Metal Furring Installation.   Metal furring should be installed as appropriate to support the acoustical wall panels or baffles. It should be placed on lines and levels necessary to cause the acoustical panels to fall into the proper location, and to properly support baffle hangers at the correct locations. Furring that will support directly applied panels or baffles should correct unevenness in a supporting structure or solid substrate.

When placed over a solid substrate, metal furring should be installed over a layer of polyethylene film or other vapor-retarding material as an airflow barrier and vapor retarder.

Furring should form a complete system adequate to properly support the acoustical panels or baffles. Closers should be installed at edges and openings.

Screws, bolts, clips, wires, and other anchors should be used to anchor the furring in place. Generally, fasteners should be placed near the ends of furring members and not more than 24 inches on center between, but other support spacings may be found. Some types of furring members may require closer spacings. The correct spacing should be verified with the furring member manufacturer.

Furring spacings may vary in practice with the substrate and the acoustical panels or baffles being supported. Minimum spacings for wall panels are dictated by the panel thickness and size, and may vary somewhat from manufacturer to manufacturer and system to system. Spacing must be such as to adequately support the panels. Spacing of furring for baffle hangers should be such that the baffle hangers fall in the correct locations.

Spline-mounted Panels.   The usual method of mounting spline-mounted acoustical wall panels is to place splines vertically and fasten them to a supported substrate, such as plaster or gypsum board (Fig. 3-23). Then the panels are slipped in place so as to engage the splines in the kerfed and rabbeted panel edges. Spline-mounted panels can also, however, be mounted by fastening the splines to wood or metal furring over masonry, concrete, or another solid substrate (see Figs. 3-21 and 3-22).

Spline-mounted acoustical wall panels may rest directly on the floor or be supported at the bottom by a base retainer, such as a J- or angle molding (see Fig. 3-18).

Panels should extend to the ceiling, but not into the ceiling plenum. Sometimes, the panels are stopped short of the ceiling leaving a decorative reveal.

When panels must be cut for fitting or other reasons, the panel backs

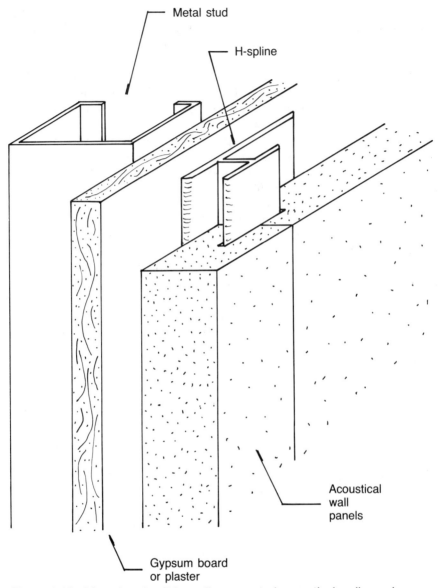

Figure 3-23 Mounting detail for spline-mounted acoustical wall panels over a supported substrate.

should be resealed using heavy mastic tape or foil to prevent airflow through the panel. Cut edges should be covered by a molding or wrapped with fabric.

Exposed panel edges should be covered with fabric or a molding.

Back-mounted panels. Back-mounted panels are usually mounted directly on a supported substrate, such as plaster or gypsum board. Most older panels and many newer panels will have been mounted using a conventional Z-clip, continuous Z-clip, or some similar proprietary device. Some will have been mounted using adhesives. Others will be supported by magnetic pads. Some will have been screw-attached or supported by battens. Many modern installations are supported using hook and loop (Velcro) strips.

Metal Panels. Many metal panels are installed using clips or screws. Some are installed using accessories similar to the moldings shown in Figure 3-18, or the clips and magnetic pads shown in Figure 3-20.

Baffles. Most baffles are suspended vertically by wire, cables, chains, bars, or fishing line from framing or furring. They should be installed in line and vertically or at the proper angle, according to the design. Baffles that are suspended in other configurations should be installed according to the design. In every case, hangers should be securely fastened to the supports and to the baffles.

Sound Insulation. Where they are used, sound insulation blankets should be installed over the full area occupied by the acoustical wall panels.

Sound insulation pads in metal wall panels should be fitted snugly into each pan, panel, or strip over the entire area of the wall panel installation.

Insulation should be tightly butted with joints in solid contact and without voids. It should be fitted neatly to adjacent surfaces.

Why Acoustical Wall Panels and Baffles Fail

Some acoustical wall panel and baffle failures can be traced to one or several of the following sources: structure failure; structure movement; solid substrate problems; supported substrate problems; other building element problems. Those sources are discussed in Chapter 2. Many of them, while perhaps not the most probable causes of acoustical wall panel or baffle failure, are more serious and costly to fix than the types of problems discussed in the following paragraphs. Consequently, the possibility that they are responsible for an acoustical wall panel or baffle failure should be investigated.

The causes discussed in Chapter 2 should be ruled out or, if found to be at fault, rectified before acoustical wall panel or baffle repairs are attempted. It will do no good to repair existing, or install new, acoustical wall panels or baffles when a failed, uncorrected, or unaccounted-for problem of the types discussed in Chapter 2 exists. The new installation will also fail. After the problems discussed in Chapter 2 have been investigated and found to be not present, or repaired if found, the next step is to discover any additional causes for the acoustical wall panel or baffle failure and correct them. Possible additional causes include bad materials, improper design, bad workmanship, failing to protect the installation, and poor maintenance procedures. The following paragraphs discuss those additional causes. Included with each cause is a numbered list of errors and situations that can cause acoustical wall panel failure. Refer to "Evidence of Failure" later in this chapter for a listing of the types of failure to which the numbered failure causes apply.

Bad Materials.   Improperly manufactured furring, accessories, or acoustical wall panels or baffles arriving at a construction site is certainly not unheard of, and that possibility should be considered when acoustical wall panels or baffles fail. Since most imperfections can be easily seen, bad materials should be eliminated before they are installed. At any rate, the number of incidents involving bad materials is small compared with the cases of bad design and workmanship.

Types of manufacturing defects that might occur include:

1. Acoustical wall panels or baffles that are not true to shape. Examples include square panels that are actually trapezoids and panels that are warped.
2. Imperfectly manufactured core or facing materials, or complete panels or baffles.
3. Facing materials that are stained or otherwise discolored at the factory.
4. Acoustical panels or baffles that are damaged in the manufacturing process.
5. Furring or accessories that are twisted, bent, bowed, warped, or otherwise malformed.

Improper Design.   Improper design includes selecting the wrong furring system or acoustical wall panel or baffle for the location and conditions, as well as requiring an improper installation.

General Problems.   The following general design problems can lead to acoustical wall panel or baffle failure.

1. Selecting acoustical wall panels or baffles with inappropriate core, facing, or edges for the location and use intended. Selecting units without impact-resistant faces for heavy-use areas where they are accessible to people is not a good idea, for example.
2. Selecting acoustical wall panels or baffles that have an inappropriate acoustical performance rating. Using a panel with a low NRC rating in a high-noise area is an example.
3. Selecting accessories that are of the wrong type for the installation, or are too small, too weak, or too few to perform the necessary function.
4. Failing to require proper installation of acoustical wall panels or baffles, including requiring that too few supports or fasteners or too little adhesive be used, or that fasteners, supports, or adhesives be placed in the wrong locations, that adhesive be improperly applied to panels, or that panels or baffles be improperly installed after the accessories are in place or the adhesive has been applied.

Wood Furring Problems.   Wood furring may fail because of structure failure, structure movement, solid or supported substrate failure, or other building element problems discussed in Chapter 2, or because of problems inherent in the furring itself.

Furring design problems that can cause acoustical wall panel or baffle failures include the following:

1. Requiring the use of flexible or extremely hard wood for furring.
2. Permitting the use of wet lumber.
3. Requiring that furring members be placed too far apart.
4. Requiring that wood furring be installed so that loads from a deflecting or otherwise moving structure pass into the furring.

Metal Furring Problems.   Metal furring may fail because of structure failure, structure movement, solid or supported substrate failure, or other building element problems discussed in Chapter 2, or because of problems inherent in the furring itself.

Furring design problems that can cause acoustical wall panel or baffle failures include the following:

1. Requiring that furring members be placed too far apart or at the wrong spacing.
2. Requiring that metal furring be installed so that loads from a deflecting or otherwise moving structure pass into the furring.

Bad Workmanship.   Correct installation is essential if acoustical wall panel and baffle failures are to be prevented.

General Problems.   The following workmanship problems can lead to acoustical wall panel and baffle failure.

1. Failing to follow the design and the recommendations of the manufacturer and recognized authorities.
2. Failing to properly prepare the area where acoustical wall panels or baffles will be installed, removing contaminants and items that will interfere with proper installation or damage the panels or baffles.
3. Failing to properly seal the back of acoustical wall panels that have been field-cut or on which the back surface or foil covering has been damaged.
4. Failing to clean the back of acoustical wall panels that will be installed using adhesives and to prime the substrate before such installation.
5. Installing acoustical wall panels or baffles before the building has been completely closed and before wet work—such as concrete, masonry, and plaster—has dried sufficiently and, when the panels are to be installed using adhesives, is at the proper level of alkalinity.
6. Installing damaged acoustical panels, baffles, furring, or accessories, regardless of whether the damage was inherent in the manufactured materials or occurred during shipment, storage, or installation.
7. Using fasteners, accessories, hangers, and other materials that will rust where high humidity will occur, or where they will be set in, or be in direct contact with, masonry or concrete.
8. Using fasteners, hangers, or accessories that are of the wrong type for the installation, or are too small or weak to perform the necessary function.
9. Failing to allow acoustical panels and baffles to acclimate to the temperature and humidity in the space before installing them, which can lead to appearance changes in the installed panels or baffles.
10. Permitting acoustical wall panels or baffles to become wet or damp before or during installation, which can cause discolorations due to water marks and mildew, or otherwise damage materials or contribute to the oxidization of metal materials.
11. Installing acoustical wall panels or baffles that are watermarked or otherwise stained.
12. Installing acoustical wall panels or baffles that have damaged edges or faces.
13. Failing to install an airflow barrier and vapor retarder, such as polyethylene film, over masonry and concrete before installing acoustical wall panels.
14. Failing to properly install acoustical wall panels or baffles using fas-

teners, hangers, supporting accessories, or adhesives, including not using enough fasteners, hangers, supporting accessories, or adhesive; placing fasteners, hangers, or accessories in the wrong locations; improperly applying adhesive to panels; and improperly installing panels after the fasteners, hangers, accessories, or adhesive have been applied.

15. Failing to install acoustical wall panels or baffles so that they finish in the proper plane within acceptable tolerances.
16. Installing acoustical wall panels or their supporting accessories, using adhesives, when the room air temperature exceeds 100 degrees Fahrenheit, or when the room or adhesive temperature is less than 50 degrees Fahrenheit.
17. Applying acoustical wall panels or their supporting accessories, using an adhesive, to damp plaster or concrete or in a damp room.
18. Applying acoustical wall panels or their supporting accessories to loose paint using an adhesive.
19. Failing to install acoustical materials behind metal wall panels.
20. Installing acoustical wall panels across an expansion or control joint in the substrate.

Wood Furring Problems.   Acoustical wall panels or baffles may fail because of problems inherent in wood furring, including:

1. Installing furring so that it is misaligned, twisted, or protruding beyond the plane in which it should be.
2. Lumber shrinkage. Even relatively dry wood will shrink.
3. Using wet lumber. When it dries, lumber that is too wet when installed can warp, split, twist, or otherwise change in shape, or may shrink away from fasteners and become loose or fall away from its supports.
4. Using flexible or extremely hard wood for furring. Wood that is too hard or too flexible may prevent fasteners from entering and seating properly, resulting in loose acoustical wall panels.
5. Using wood that is too thin for its span.
6. Installing furring members too far apart or at the wrong spacing.
7. Failing to properly support and anchor furring members. When the furring loses stability or separates from the substrate, supported acoustical wall panel failure is inevitable.
8. Leaving out portions of the furring, such as at openings and edges, so that there is nothing to which the splines may be fastened.
9. Installing wood furring so that loads from a deflecting or otherwise moving structure pass into the furring. Examples include wedging the ends of furring against masonry or concrete and building the ends of furring members into masonry.

10. Extending wood furring across an expansion or control joint in the substrate.

Metal Furring Problems. Acoustical wall panels or baffles may fail because of problems inherent in metal furring, including the following.

1. Installing furring so that it is misaligned, twisted, or protruding beyond the plane in which it should be.
2. Placing furring members too far apart or at the wrong spacing.
3. Failing to properly support and anchor furring members. When the furring loses stability or separates from the substrate, supported acoustical wall panel failure is inevitable.
4. Leaving out portions of the furring, such as at openings and edges, so that there is nothing to which the splines may be fastened.
5. Installing metal furring so that loads from a deflecting or otherwise moving structure pass into the furring. Examples include wedging the ends of furring against masonry or concrete, and building the ends of furring members into masonry.
6. Extending metal furring across an expansion or control joint in the substrate.

Failing to Protect the Installation.   Acoustical wall panels or baffles must be protected before, during, and after installation. Errors include the following.

1. Failing to protect acoustical wall panels or baffles from staining or other damage by other construction materials and procedures.
2. Permitting abuse before or during installation, or after the acoustical wall panels or baffles have been installed.

Poor Maintenance Procedures.   Not maintaining acoustical wall panels or baffles properly can result in failures. Poor maintenance procedures incude:

1. Failing to clean panels or baffles regularly. Unremoved grease, dirt, and other contaminants can result in permanent stains.
2. Improper cleaning of acoustical wall panels or baffles. Using improper chemicals and bleaches may harm fabrics. The manufacturer's instructions should be followed.

Evidence of Failure

In the following paragraphs, acoustical wall panel or baffle failures are divided into failure types, such as "Stained or Discolored Acoustical Wall Panels or Baffles." Following each failure type are one or more failure sources, such as "Improper Design: General Problems." After each failure source, one or more numbers is listed. The numbers represent possible errors associated with that failure source that might cause that failure type to occur.

A description and discussion of the numbered failure causes for failure types "Steel and Concrete Structure Failure," "Wood Structure Failure," and "Structure Movement" appear under those headings in Chapter 2, under the main heading "Structural Framing Systems," and are listed in the Contents.

A description and discussion of the numbered failure causes for failure types "Solid Substrate Problems" and "Supported Substrate Problems" appear under those headings in Chapter 2, under the main heading "Substrates." They are listed in the Contents.

A description and discussion of the numbered failure causes for failure type: "Other Building Element Problems" appears under that heading in Chapter 2, under the main heading "Other Building Elements," and is listed in the Contents.

The following example applies to the failure types listed in the preceding three paragraphs. Clarification and explanation of the numbered cause (1) in the example

■ Supported Substrate Problems: 1 (see Chapter 2).

appears in Chapter 2 under the main heading "Substrates" subheading "Supported Substrate Problems," Cause 1, which reads

1. Loose Plaster.

A description and discussion of the types of problems and numbered failure causes that follow failure types "Bad Materials," "Improper Design," "Bad Workmanship," "Failing to Protect the Installation," and "Poor Maintenance Procedures" appear under those headings earlier in this chapter, under the main heading "Acoustical Wall Panels and Baffles," subheading "Why Acoustical Wall Panels and Baffles Fail," which are listed in the Contents. For example, clarification and explanation of the type of problem (Bad Workmanship: General Problems) and numbered cause (12) in the example

■ Bad Workmanship: General Problems: 12.

appears earlier in this chapter under the main heading "Acoustical Wall Panels and Baffles" subheading "Why Acoustical Wall Panels and Baffles Fail," sub-subheading "Bad Workmanship," sub-sub-subheading "General Problems," Cause 12, which reads:

12. Installing acoustical wall panels or baffles that have damaged edges or faces.

Poor Acoustical Performance.   Every acoustical wall panel and baffle failure discussed in this chapter can affect acoustical performance. In addition, panels and baffles may exhibit unsatisfactory acoustical performance for one of the following reasons.

- Other Building Element Problems: 1, 2 (see Chapter 2).
- Bad Materials: 2, 4.
- Improper Design: General Problems: 2.
- Bad Workmanship: General Problems: 1, 5, 6, 10, 12, 19.

Stained or Discolored Acoustical Wall Panels or Baffles.   Blemishes include watermarks, mildew, and other stains and discolorations. Any one or more of the following may be at fault.

- Solid Substrate Problems: 1 (see Chapter 2).
- Supported Substrate Problems: 5, 6, 13 (see Chapter 2).
- Other Building Element Problems: 1, 2 (see Chapter 2).
- Bad Materials: 2, 3, 4.
- Improper Design: General Problems: 1.
- Improper Design: Wood Furring Problems: 2.
- Bad Workmanship: General Problems: 1, 2, 3, 5, 6, 7, 10, 11, 13.
- Bad Workmanship: Wood Furring Problems: 3.
- Failing to Protect the Installation: 1, 2.
- Poor Maintenance Procedures: 1, 2.

Damaged Acoustical Wall Panels and Baffles.   Damage may be cracked or split panels or baffles, crushed or broken panel or baffle edges, depressions in panel faces, holes through panels or baffles, torn or split facing material. Any one or several of the following may be at fault, depending on the type and extent of the damage.

- Steel and Concrete Structure Failure: 1 (see Chapter 2).
- Wood Structure Failure: 1, 2, 3, 5, 6, 8 (see Chapter 2).
- Structure Movement: 1, 2, 3, 4, 5, 7, 8 (see Chapter 2).
- Solid Substrate Problems: 2, 3, 4 (see Chapter 2).
- Supported Substrate Problems: 3, 9, 11, 12 (see Chapter 2).
- Bad Materials: 2, 4, 5.
- Improper Design: General Problems: 1, 4.
- Improper Design: Wood Furring Problems: 4.
- Improper Design: Metal Furring Problems: 2.
- Bad Workmanship: General Problems: 1, 2, 6, 8, 9, 12, 14 20.

- Bad Workmanship: Wood Furring Problems: 1, 2, 3, 7, 9, 10.
- Bad Workmanship: Metal Furring Problems: 1, 3, 5, 6.
- Failing to Protect the Installation: 1, 2.
- Poor Maintenance Procedures: 2.

Loose, Sagging, Fallen, Raised, or Out-of-line Acoustical Wall Panels or Baffles. Acoustical wall panels or baffles may become loose, sag, separate from a substrate or furring, or be out of alignment with other units due to one or more of the following causes.

- Steel and Concrete Structure Failure: 1 (see Chapter 2).
- Wood Structure Failure: 1, 2, 3, 4, 5, 6, 7, 8 (see Chapter 2).
- Structure Movement: 1, 2, 3, 4, 5, 6, 7, 8 (see Chapter 2).
- Solid Substrate Problems: 1, 2, 3, 4 (see Chapter 2).
- Supported Substrate Problems: 1, 2, 3, 4, 8, 9, 10, 11, 12, 16 (see Chapter 2).
- Other Building Element Problems: 1, 2 (see Chapter 2).
- Bad Materials: 1, 2, 4, 5.
- Improper Design: General Problems: 1, 3, 4, 5.
- Improper Design: Wood Furring Problems: 1, 2, 3, 4.
- Improper Design: Metal Furring Problems: 1, 2.
- Bad Workmanship: General Problems: 1, 2, 3, 4, 5, 6, 7, 8, 9, 10, 14, 15, 16, 17, 18, 20.
- Bad Workmanship: Wood Furring Problems: 1, 2, 3, 4, 5, 6, 7, 8, 9, 10.
- Bad Workmanship: Metal Furring Problems: 1, 2, 3, 4, 5, 6.
- Failing to Protect the Installation: 2.

Cleaning, Repairing, Refinishing, and Extending Acoustical Wall Panels and Baffles

General Requirements. The following paragraphs contain some suggestions for cleaning, repairing, refinishing, and extending acoustical wall panels and baffles and associated furring and suspension systems. Because the suggestions are meant to apply to many situations, they might not apply to a specific case. In addition, there are many possible cases that are not specifically covered here. When a condition arises in the field that is not addressed here, advice should be sought from the additional data sources mentioned in this book. Often, consultation with the manufacturer of the materials being repaired will help. Sometimes, it is necessary to obtain professional help (see Chapter 1). Under no circumstances should the specific recommendations in this chapter be followed without careful investigation and application of professional expertise and judgment.

Before an attempt is made to clean, repair, refinish, or extend existing acoustical wall panels and baffles, the existing materials manufacturers' brochures for products, installation details, and recommendations for cleaning and refinishing should be available and referenced. It is necessary to be sure that the manufacturer's recommended precautions against materials and methods which may be detrimental to finishes and acoustical efficiency are followed.

Where existing acoustical wall panels and baffles to be repaired or extended are installed using staples, nails, screws, or other fasteners, or are installed using adhesives, their manufacturer's recommendations for methods to be used in the repairs or extensions should be obtained. Often, it is not possible to satisfactorily remove materials installed in such ways without damaging the units to the extent that they are not suitable for reuse. Trying to do so without following the manufacturer's instructions will almost certainly lead to failure.

A check should be made to see whether sufficient extra material was left from the original installation to make the repairs. Often, as much as three percent of the amount of acoustical material actually installed will be left with the owner for making future repairs.

Unless existing wall panels or baffles must be removed to facilitate repairs to other portions of the building, or are scheduled to be replaced with new materials, only sufficient materials should be removed as is necessary to affect the cleaning and repairs. The more materials that are removed, the greater the possibility of additional damage occurring during handling and storage of the removed materials. Materials that are removed should include those that are loose, sagging, or damaged beyond in-place repair. Other existing wall panels, baffles, furring, or suspension components that are sound, adequate, and suitable for reuse may be left in place, or removed, cleaned, and stored for reuse.

After acoustical wall panels and baffles have been removed, concealed damage to furring or suspension system components may become apparent. Such concealed damage should be repaired before repaired, cleaned, or new acoustical wall panels or baffles are installed. Such additional damage may include furring or suspension system components that are damaged, rusted beyond repair, or otherwise unsuitable for reuse. Existing furring, hangers, and hanger attachments that are sound, adequate, and suitable for reuse may be left in place, or removed, cleaned, and stored for reuse.

Existing acoustical wall panels and baffles, and associated furring and suspension system components that are to be removed, should be removed carefully, and adjacent surfaces should be protected, so that the process does no damage to the surrounding area. Materials and components should be removed in small sections. Temporary supports should be installed when necessary to prevent the collapse of construction that is not to be removed. Removed materials that will be reinstalled should be handled carefully, stored safely, and protected from damage. Debris should be removed promptly, so that it will not be responsible for damage to materials that are to remain in place.

Unless the decision is made to discard them, damaged acoustical wall panels, baffles, furring, and suspension system components that can be satisfactorily repaired should be repaired, whether they have been removed or left in place. A failure to repair known damage may lead to additional failure later on. Methods recommended by the manufacturer of the materials should be followed carefully when making repairs, even when the repairs are as simple as touch-up painting. Acoustical materials, furring, and suspension system components that cannot be satisfactorily repaired should be discarded and new matching units installed.

Areas where repairs will be made should be inspected carefully to verify that existing components that should be removed have been removed, and that the substrates and structure are as expected, and are not damaged. Sometimes, substrate or structure materials, systems, or conditions are encountered which differ considerably from those expected. Sometimes unexpected damage is discovered. Both damage that was previously known and damage found later should be repaired before acoustical wall panels or baffles are reinstalled. Acoustical wall panel or baffle reinstallation should not proceed until other work to be concealed or made inaccessible by the panels or baffles has been completed, and unsatisfactory conditions have been corrected.

With a few exceptions, removed acoustical materials, furring, and suspension system components that are in an acceptable condition may be reinstalled. Where possible, they should be reinstalled in the same location from where they were removed. Materials and components that are installed in locations other than those from which they were removed may not match the surrounding materials, which will make patches obvious. An exception, of course, would be the installation of removed wall panels or baffles in another space, where the removed materials are sufficient in quantity to make up the entire new installation.

Removed hangers and hanger attachments should not be reused unless they are in exceptionally good condition. Metal fatigue or other difficult-to-detect damage may exist, which may become apparent only when the reused hanger or hanger attachment fails.

It may not be possible to reinstall acoustical materials that were installed using staples, screws, nails, or other similar fasteners. The removing process will often damage them too severely.

Unless a decision is made to leave a substrate exposed that was exposed by removing all or part of an existing acoustical wall panel installation, or to install another wall panel system, salvaged acoustical wall panels should be reinstalled in accordance with the manufacturer's recommendations. Salvaged baffles should also be reinstalled in accordance with the manufacturer's recommendations. Where acceptable existing materials are insufficient in quantity, new matching materials should be used to complete the installation. During the process, gaps and openings in the existing wall panel installation, including those that existed before the repairs began and those resulting from discarding unrepairable materials, should be filled with new acoustical material that matches the existing materials. Similarly, new matching baffles should be installed where baffles are missing.

Where patching or extending existing acoustical wall panels or baffles is required, and acoustical materials or suspension systems do not exist, or existing materials are not acceptable for reuse, or existing materials are in insufficient quantity to complete the work, new acoustical materials and suspension systems should be provided. The new acoustical materials and suspension systems should match the existing adjacent undisturbed (or originally installed and removed) material in type, size, material, edge condition, thickness, pattern, finish, and characteristics. The new suspension system, or installation method, as applicable, should exactly match, as applicable, that existing in similar installations, or used in the original removed panels or baffles, or still existing in adjacent panels or baffles. Substrates should be repaired and cleaned as necessary to obtain a satisfactory installation.

Patched or extended wall panel installations should be complete, with no voids or openings, should produce a finished wall in the same plane as the existing wall panels that were left in place, and should have joints and trim of the same type as those in the existing wall in every respect. Patches and extensions should be made as inconspicuous as possible.

Damaged acoustical materials and suspension systems should not be reinstalled. Minor damage may be repaired, however, and the repaired components installed if satisfactory results can be obtained.

Removed ceiling, base, corner, and joint trim should be reinstalled in the same locations from where they were removed. New matching trim should be installed where the existing trim is damaged and where pieces of existing trim are missing.

Care should be exercised to ensure that the integrity of fire-rated systems is maintained. Acoustical materials, furring, and suspension system materials and installation should be identical with applicable assemblies that have been tested and listed by recognized authorities and are in compliance with the requirements of the building code. Materials for use in existing fire-rated assemblies should, unless doing so violates the previous sentence, exactly match the materials in the existing fire-rated assembly.

Wood Framing and Furring. Repair of major wood framing members, such as joists, rafters, trusses, truss-joists, beams, and girders is beyond the scope of this book. The following paragraphs contain only enough data to help in recognizing that those elements may be at fault when acoustical wall panels fail, and to present a general indication of what steps might be necessary to solve the problem.

Most failed wood framing or furring members are removed and new material installed, because (even when doing so is possible) it is not often reasonable to repair damaged wood framing or furring members. Straightening a warped or twisted joist, for example, will probably prove to be impracticable. So, most of the references in the following paragraphs to repairing framing or furring mean removing the damaged pieces and installing new pieces.

The following paragraphs assume that damaged concrete or metal structure and solid or supported substrate supporting wood furring has been repaired and presents a satisfactory support system for the furring being repaired.

Materials. Materials used to repair existing wood framing or furring should match those in place as nearly as possible, but should not be lesser in quality, size, or type than those recommended by recognized authorities or required by the building code. Where the existing materials are fire-retardant treated, the new materials must be similarly treated. Where the existing material is pressure preservative treated, the new should be also.

It is usually best to match lumber sizes exactly when installing a new member in an existing framing or furring system. In older buildings, however, the existing lumber may be of sizes that are no longer standard. Members that are nominally 2 by 4 may be actually 2 by 4 inches if the building is old, 1-5/8 by 3-5/8 inches, as was once the standard, or 1-1/2 by 3-1/2 inches, as is the current standard. When lumber of the exact size to match that existing is not available, there are several possible solutions to the problem. Larger members can be cut down to the size of the existing members. This may be expensive, however. For furring, it will usually be less expensive to shim the new members so that the face surfaces are in alignment with the faces of the existing members. Another alternative is to build up new members from two standard-sized members.

Standards. In general, repairs should be made in accordance with the recommendations of recognized standards, such as those mentioned earlier in this chapter or in "Where to Get More Information" at the end of this chapter, or both, and the standards referenced in the Bibliography as applicable to this chapter.

Preparation.   Where existing framing or furring members are damaged, the acoustical wall panels must be removed to the extent necessary to permit the repairs to be made. Then the damaged existing framing or furring can be removed. Damaged elements that should be removed include wood members that are twisted, warped, broken, rotted, wet, out of alignment, or otherwise unsuitable for use.

Misaligned, Warped, or Twisted Framing or Furring.   Even when so misaligned, twisted, or warped that it causes damage to applied wall panels, wood framing or furring is often left in place and the acoustical wall panels reattached in such a way that the poor condition of the framing or furring is overcome. When repair to the acoustical wall panels alone will not prevent failure from recurring, it will be necessary to remove the damaged framing or furring and provide new materials.

Shrinkage.   Where acoustical wall panel failure is caused by shrinkage in the lumber used in framing or furring, it is often possible to repair and reattach the wall panels without removing the lumber. Where wall panel repairs cannot be accomplished satisfactorily, it may be possible to remove a portion of the wall panels and shim the framing or furring to produce a flush surface for wall panel application. Where shimming does not solve the problem, it may be necessary to remove the damaged lumber and provide new framing or furring.

Incorrectly Constructed Framing or Furring.   Where the framing or furring was originally built in such a way that the acoustical wall panels become damaged, the necessary corrective measures depend on the type of error. The possibilities are many, and the solutions even more numerous. They range from simply planing down a projecting brace to reconstructing a section of wall framing.

Each case must be examined to determine the true cause before steps are taken to correct supposed errors. The extent of the damage must be also taken into account. If it might be possible to prevent the failure from recurring by simply refastening the acoustical wall panels, for example, that should be done before expensive reconstruction of the framing or furring is undertaken.

Failed Wall Panel Furring.   Sagging, bowing, or misalignment of acoustical wall panels that are directly applied to wood furring is often due to failure of the furring.

Wood furring for acoustical wall panels may fail for one of several reasons, including structure deflection; differential movement in the structure; or improper design or installation of the furring.

When acoustical wall panel failure is due to wood furring failure, the wall panels must be removed to the extent necessary to make repairs, before repairs can be made. If the acoustical wall panels have been applied using fasteners or adhesives, it may be necessary to discard the removed acoustical wall panels and provide new panels.

Metal Framing and Furring.   While failed stud wall or partition framing may be responsible for acoustical wall panel failure, repairing it is beyond the scope of this book. It should be investigated as a possible source of the failure and repaired as necessary.

The following paragraphs assume that damaged concrete or metal structure and solid or supported substrates supporting metal furring have been repaired and presents a satisfactory support system for the furring being repaired.

Most failed metal furring members are removed and new material installed, because (even when doing so is possible) it is not often reasonable to repair damaged metal furring members. So, most of the references in the following paragraphs to repairing furring mean removing the damaged pieces and installing new pieces.

Materials.   Materials used to repair existing metal furring should match those in place as nearly as possible but should not be lesser in quality, size, or type than those recommended by recognized authorities or required by the building code.

Standards.   In general, repairs should be made in accordance with the recommendations of recognized standards.

Preparation.   Where existing furring members are damaged, the acoustical wall panels must be removed to the extent necessary to permit the repairs to be made. Then the damaged existing furring can be removed. Damaged elements that should be removed include members that are twisted, warped, out of alignment, or otherwise unsuitable for use.

Misaligned, Warped, or Twisted Furring.   When so misaligned, twisted, or warped that it causes damage to applied acoustical wall panels, metal furring is usually removed and new furring provided. Where misalignment alone is the problem, however, reinstallation will often offer a solution.

Incorrectly Constructed Framing or Furring.   Where the framing or furring was originally built in such a way that the acoustical wall panels become damaged, the necessary corrective measures depend on the type of error. The possibilities are many, and the solutions even more numerous.

Each case must be examined to determine the true cause before steps are taken to correct supposed errors. The extent of the damage must be also taken into account. If it might be possible to prevent the failure from recurring by simply refastening the acoustical wall panels, for example, that should be done before expensive reconstruction of the framing or furring is undertaken.

Failed Wall Furring. Sagging, bowing, or misalignment of acoustical wall panels that are directly applied to metal furring is often due to failure of the furring.

Metal furring for acoustical wall panels fails for several reasons, including structure deflection; differential movement in the structure; or improper design or installation of the furring.

When acoustical wall panel failure is due to metal furring failure, the wall panels must be removed to the extent necessary to make repairs. If the acoustical wall panels have been applied using fasteners or adhesives, it may be necessary to discard the removed acoustical wall panels and provide new panels.

Rusted Furring. Members that are extensively rusted are probably not worth saving. Minor rust should be removed and the surface recoated with the material originally used. Painted members should be repainted with a compatible paint. Galvanized members should be repaired using galvanizing repair paint.

Baffle Hangers. It may not be feasible to repair failed baffle hangers or their attachments. So, much of the time when this text refers to repairing metal hangers and their attachments, it means removing the damaged pieces and installing new pieces in their place.

The following paragraphs assume that structure and solid substrate damage has been repaired and that there is satisfactory support for the hangers and attachments being repaired.

Materials. Components used to repair existing metal baffle hangers and hanger attachments should match those in place as nearly as possible, but should not be lesser in quality, size, or type than those recommended by the baffle manufacturer or required by the building code.

Damaged existing materials should not be used in making repairs or for reinstalling existing baffles or installing new baffles unless they can be satisfactorily repaired.

Hangers and hanger attachments left in place when baffles are removed should be examined carefully and their adequacy and suitability verified. Those found to be unsuitable should be removed.

When installing new baffles or reinstalling removed existing baffles, left-in-place hangers and hanger attachments may be used where they are suitable. New hangers and hanger attachments must be provided where existing hangers and attachments are removed, where existing hangers or attachments are improperly placed or otherwise inappropriate, and where there are no existing hangers. New hangers should be attached to concrete, structural steel, or wood framing members. They should never be attached to steel deck, pipes, conduits, ducts, or mechanical or electrical devices, and existing hangers that are so attached should not be used.

New hangers should be plumb and not in contact with other elements, such as insulation covering ducts and pipes. Hangers should be splayed only where obstructions or conditions prevent plumb, vertical installation. The horizontal forces generated by splaying of hangers should be offset by counter-splaying, bracing, or another suitable method.

Standards.   In general, repairs should be made in accordance with the recommendations of the baffle manufacturer.

Preparation.   Where existing hangers and hanger attachments are damaged, the supported acoustical baffles must be removed to the extent necessary to permit repairs to be made. Then the damaged hangers and attachments can be removed. Removed damaged elements that are not repairable should be discarded. Damaged elements that should be removed include, but are not limited to, hangers and attachments that are broken, bent, or otherwise physically damaged, rusted beyond repair, or otherwise unsuitable for reuse. Existing hangers and hanger attachments that are sound, adequate, and suitable for reuse may be left in place, or removed, cleaned, and reused.

Hangers and hanger attachments that have been left in place where baffles have been removed should be examined to verify their adequacy and suitability for use in rehanging removed baffles or hanging new baffles. Those found to be unsuitable should be removed.

Existing hangers and hanger attachments in good condition may be used to support new or reinstalled existing baffles. New hangers and attachments should be installed, of course, where existing hangers and attachments have been removed or are improperly placed or otherwise inappropriate, and where there are no existing hangers. New hangers should be attached to the structure, never to pipes, conduits, ducts, mechanical or electrical devices, or steel deck. Existing hangers so attached should not be used.

Failed Hangers and Hanger Attachments.   Hangers and their attachments may fail for several reasons, including improper design or installation.

When an acoustical baffle failure is due to hanger or hanger attachment failure, it will probably be necessary to remove the baffles in order to repair the hangers or attachments.

Acoustical Wall Panels and Baffles

Materials.   Acoustical wall panels for use in patching and extending existing acoustical wall panels should exactly match the existing materials. New baffles should also exactly match adjacent baffles that are to remain in place. Panels or baffles other than the same brand used in the original installation will almost certainly not match the panels originally used.

Cleaning.   The manufacturer's instructions should be obtained and followed when cleaning acoustical wall panels or baffles. The following suggestions are general in nature and should not take precedence over the manufacturer's instructions.

Panels and baffles should be vacuumed regularly to prevent dust and dirt build-up.

Some fabric-covered panel and baffle surfaces can be cleaned with a wet or dry shampoo of the type formulated for use on upholstery. Dry shampoos are available in aerosol cans. They spray on as a foam, which is then worked into the fabric with a dry sponge and vacuumed off to remove the shampoo and suspended dirt.

Wet shampoos are also available in spray cans. They spray on as a liquid that turns to foam after it is sprayed. The foam is then worked into the fabric and vacuumed off.

Solvent cleaners may be used on some fabric-covered wall panels and baffles to remove oil or grease stains that do not respond to shampoos.

Some fabric-covered panels and baffles that have become badly stained or soiled may be steam cleaned using commercial upholstery cleaners.

Metal-covered panels and baffles may be cleaned by wiping with a damp cloth. They may also be washed with soap and water if the acoustical pads are first removed. Painted or otherwise finished metal panels may be cleaned or refinished in the same manner as that suggested for metal acoustical ceiling materials in the ''Acoustical Ceilings'' part of this chapter.

The more severe the cleaning method used, the more care that must be exercised. Shampoos are relatively benign on most fabric-covered panels and baffles. Steam cleaning may destroy some materials. Water cannot be used to clean some panels and baffles. Perforated metal panels with acoustical pads are particularly susceptible to damage by water. In every case, the panel or baffle manufacturer's instructions must be followed exactly.

Surface Damage. Torn or abraded fabric covering on wall panels and baffles may be repaired if the damage is not too severe. Most repairs will be visible, but some may not be too objectionable. Panels and baffles that are damaged too severely to be satisfactorily repaired must be removed and new material installed. Sometimes, it will be possible to recover damaged panels with matching fabric. All such repairs should be done by professionals, following the directions of the panel or baffle manufacturer.

Minor scratches, nicks, and chips in painted wall panels and baffles may be concealed using chalk, pastel, typist correction fluid, shoe polish, or a small amount of paint.

Most painted acoustical wall panels and baffles can be repainted without significant loss of acoustical proficiency. The manufacturer should be consulted, however, before proceeding. The panel or baffle manufacturer may have specific recommendations regarding which paint material to use. The gloss of the paint should, of course, match the gloss of the original finish.

Paint may be applied using either spray, roller, or brush, as the panel or baffle manufacturer recommends. However, some surface patterns may significantly lose acoustical performance when spray painted. It may be necessary to remove acoustical pads from metal panels and baffles before painting, especially if the paint is sprayed on.

Regardless of the paint material and application method used, the applicator must be careful to ensure that openings in the panels or baffles that are required for acoustical performance are not clogged with paint. Blocking the perforations will cause a significant loss of acoustical performance.

Acoustical wall panels and baffles to be painted should be thoroughly cleaned of dust, dirt, and other foreign particles, and permitted to dry completely. One coat is normally sufficient, if the selected paint has good hiding qualities, but two coats may be necessary in some cases, especially if it is necessary to thin the paint to prevent sealing the perforations in the acoustical panels or baffles.

Clear coatings on natural aluminum finishes can usually be repaired using the same clear coating originally used, but such repairs may be visible. Such clear coatings are usually, but not always, lacquer.

Repair of anodized aluminum, plastic coatings, and fluoropolymer finishes should be made in accordance with the panel or baffle manufacturer's and finisher's recommendations. Similar coatings may not respond alike to the same refinishing method. Techniques that work well with one finish may be harmful to another, similiar finish.

Acoustical Wall Panel and Baffle Installation. Removed panels and baffles should generally be placed in the same location from which they were removed, so that the direction of the pattern and the appearance of the finished installation is not altered from that of the original application. When wall panel installation has been completed, the entire wall area should be covered with acoustical panels with no gaps or openings. Where necessary, acoustical materials should be cut around outlets, penetrations, and accessories.

Installing New Acoustical Wall Panels and Baffles over Existing Materials

General Requirements. The following paragraphs contain some suggestions for installing new acoustical wall panels, including their furring, and baffles, including their hangers and hanger attachments, in existing spaces. Because the suggestions are meant to apply to many situations, they might not apply to a specific case. In addition, there are many possible cases that are not specifically covered here. When a condition arises in the field that is not addressed here, advice should be sought from the additional data sources mentioned in this book. Often, consultation with the manufacturer of the materials being installed will help. Sometimes, it is necessary to obtain professional help (see Chapter 1). Under no circumstances should these recommendations be followed without careful investigation and the application of professional expertise and judgment.

The discussion earlier in this book about new acoustical wall panel and baffle materials and installations also applies to new wall panels and baffles installed in existing spaces. There are, however, a few additional considerations to address when the new wall panels or baffles are installed in an existing space. The following paragraphs address those concerns. Some, but not all, of the requirements discussed earlier are repeated here for the sake of clarity. It is suggested that the reader refer to earlier paragraphs under this same heading ("Acoustical Wall Panels and Baffles") for additional data. Specific applicable headings include:

- Panel and Baffle Materials: "Panels and Baffles."
- Substrates, Furring, and Metal Suspension Systems: "Support Systems."
- Fire Ratings: "Fire Performance."
- Furring and Acoustical Wall Panel and Baffle Installation: "Installation."

Existing areas where new acoustical wall panels or baffles will be installed should be inspected to verify that existing materials that are to be removed have been removed, work to be concealed by the panels has been completed, the area is ready to receive the wall panels or baffles, and the actual conditions are the same as those expected.

Hangers and hanger attachments left in place when existing baffles were removed should be inspected to verify their adequacy and suitability for the support of new baffles. Those found to be unsuitable should be removed and new hangers and attachments installed in their place.

Acceptable hangers and their attachments may be used to support new baffles in existing spaces. New hangers and hanger attachments must be installed, of course, when existing hangers and attachments have been removed, where existing hangers or attachments are improperly placed or are otherwise inappropriate, and where there are no existing hangers. Removed hangers and hanger attachments should not generally be reused, unless their condition is exceptional. Such removed items may have hidden damage. Hangers should be attached to structural elements, not to steel deck, pipes, conduits, ducts, or mechanical or electrical devices. Existing hangers so attached should not be reused.

Installing New Wood Furring over Existing Materials. Where new acoustical wall panels are to be applied in an existing wood-framed space, a new wood furring system is sometimes used. Installing new, and repairing and extending existing, structural framing systems are beyond the scope of this book.

Even in new buildings, wood furring is always applied over something else. The problems vary only slightly when the something else is old. More shimming may be necessary when new furring is applied on existing substrates, of course, and sometimes an existing surface will be in such poor shape that it cannot be satisfactorily furred. When that happens, it may be necessary to remove the existing construction completely and build it new.

It is not always necessary to remove existing trim when installing a new furring system. Items that will be concealed in the finished installation can be left in place.

New wood furring can be applied directly over existing finishes that are themselves directly applied to wood studs when the existing framing is sufficiently strong to hold both the new wall panels and the existing finish. New furring strips should be laid perpendicular to the existing framing, at a spacing appropriate for the new wall panels, and nailed or screwed through the existing finish to the existing framing. The nails or screws should extend into the existing framing at least 1-1/4 inches.

Installing New Metal Furring over Existing Materials. Where new acoustical wall panels are to be applied in an existing space, a new metal furring system is sometimes used.

Even in new buildings, metal furring is always applied over something else. The problems vary only slightly when the something else is old. More shimming may be necessary when new furring is applied on existing substrates, of course, and sometimes an existing surface will be in such poor shape that it cannot be satisfactorily furred. When that happens, it may be necessary to remove the existing construction completely and build it new.

New metal furring can sometimes be applied directly onto wall finishes that are themselves directly applied to wood or metal framing or furring, when the existing framing or furring is sufficiently strong to hold both the existing finish and the new wall panels. New furring members should be laid perpendicular to the existing framing or furring, at a spacing appropirte for the new wall panels, and fastened through the existing finish to the underlying framing or furring.

Installing New Acoustical Wall Panels.   Requirements stated earlier in this chapter for all new installations in new spaces apply as well as acoustical wall panels in existing spaces.

Installing New Acoustical Baffles.   The principles discussed earlier in this chapter for installing baffles in a new building are applicable when installing baffles in an existing space.

Even in new buildings, hangers for baffles are always hung from something else. The problems vary only slightly when the something else is old.

Baffles are often installed in existing spaces that do not have suspended ceilings. In such cases, hanger attachments can often be attached directly to the existing framing, just as they would be in a new building.

Even where ceilings exist, it is not usually necessary to completely remove them when installing new hangers and hanger attachments for baffles. Access to the supporting structure must be gained, of course, but items that do not block that access can, and should, be left in place.

Where the existing framing is not at the proper location for new baffle hangers, carrying channels are sometimes introduced to achieve the proper spacing. The carrying channels are, of course, suspended from the structure and the baffle hangers are suspended from the carrying channels. The acoustical baffle manufacturer should be consulted before such special applications are attempted, however.

In every case, care must be taken to ensure that new hangers or carrying channels are attached to structural framing elements. Hangers should not be attached to steel decks, pipes, conduit, ducts, or mechanical or electrical devices.

Where to Get More Information

The Ceilings and Interior Systems Contractors Association's Pamphlet "Acoustical Ceilings—Use and Practice" is a good guide to the selection and installation of acoustical ceilings. It is available from CISCA and from some acoustical ceiling manufacturers.

The National Forest Products Association's *Manual for House Framing* contains a comprehensive nailing schedule and other significant data about wood framing. It is a useful tool for anyone who must deal with wood construction of any type.

The Forest Products Laboratory's *Handbook No. 72—Wood Handbook* contains a detailed discussion of wood shrinkage.

Ramsey/Sleeper's *Architectural Graphic Standards* contains useful data about acoustical ceilings, metal ceilings, wall panels and units, and acoustical baffles.

Some of AIA Service Corporation's *Masterspec* Basic sections contain excellent descriptions of the materials and installations that are addressed in this chapter. Unfortunately, those sections contain little that will help with troubleshooting failed acoustical treatment. Sections that have applicable data are:

- Section 06100, Rough Carpentry.
- Section 09510, Acoustical Ceilings.
- Section 09521, Acoustical Wall Panels.
- Section 13070, Integrated Ceilings.

Every designer should have a full complement of applicable ASTM Standards, of course, but anyone who needs to understand acoustical ceiling materials and their suspension systems should definitely own a copy of the following ASTM Standards:

- Standard C 635, "Metal Suspension Systems for Acoustical Tile and Lay-In Panel Ceilings."
- Standard C 636, "Installation of Metal Ceiling Suspension Systems for Acoustical Tile and Lay-In Panels."
- Standard D 1779, "Specification for Adhesive for Acoustical Materials." ASTM.
- Standard E 84, "Test Method for Surface Burning Characteristics of Building Materials." ASTM.
- Standard E 119, "Fire Tests for Building Construction and Materials." ASTM.

Other ASTM Standards that are of interest to anyone involved with acoustical treatment are listed in the Bibliography and marked with a [3]. They may also be mentioned in this and other chapters.

Also refer to other items marked [3] in the Bibliography.

4

Resilient Flooring

Resilient flooring discussed in this chapter includes tile and sheet products. Tile products discussed include vinyl; vinyl composition, including asbestos; rubber; and asphalt. Sheet products discussed include linoleum, rubber, and vinyl. Liquid-applied flooring products are beyond the scope of this book.

Before proceeding, it will be helpful to define some of the terms we will use in this chapter. When this chapter uses the following terms, the listed definitions apply.

Resilient flooring: All types of tile and sheet flooring discussed in this chapter are collectively called resilient flooring. Liquid-applied flooring is not included.

Sheet flooring: Resilient flooring materials that are furnished in rolls, usually 6 to 12 feet wide, or in sheet form.

Tile: Resilient flooring materials that are supplied and installed as small units are called tile. Tile sizes may range from 9 inches square to 24 inches square.

Vinyl: Polyvinyl chloride.

Resilient Flooring Materials

Resilient flooring materials are used for floors in commercial and residential applications. They are used in virtually every type of space, including (but certainly not limited to) lobbies, corridors, offices, residential kitchens, school rooms, apartment laundry rooms, and elevators. Resilient flooring is also used on elevated access flooring in places such as laboratories and computer rooms.

Characteristics Common to All Types of Resilient Flooring

Colors and Patterns. Resilient flooring products are available in many colors and patterns. Determining which color or pattern was used in an existing installation may be impossible, unless the exact product used can be ascertained.

Fire Resistance. Though not always the case, the fire resistance of a particular resilient floor may have been dictated by a code or insurance requirement. It is imperative that any such requirements be determined, and that new materials used to repair or replace existing resilient flooring have the same fire resistance characteristics as the existing material.

Significant fire resistance characteristics of resilient flooring include flame spread and critical radiant flux (CRF). ASTM standard E 84 is sometimes noted as the test used for rating a particular resilient flooring, but may not be appropriate and may not comply with some codes because of the methods used in that test. The test method which is stipulated in ASTM standard E 648 or NFPA 253 is a more appropriate method, and is the test referenced in the BOCA Basic Building Code and the Life Safety Code (NFPA 101).

Ratings for most installations should be as follows:

Critical radiant flux (CRF) is usually 0.45 watts per square centimeter when rated in accordance with ASTM Standard E 648. Some codes may require lower values, however.

Flame spread should be not more than 75 when tested in accordance with ASTM Standard E 84.

Smoke developed should be not more than 450 when tested in accordance with ASTM Standard E 84.

Smoke density should be not more than 450 when tested in accordance with ASTM Standard E 662.

Resilient Floor Material Selection. Several factors should be considered when selecting resilient flooring. The length of service and damage that appears later can be greatly affected by whether the correct criteria were addressed in making the original choice. Criteria that should be considered when making the initial selection and that may influence resilient flooring life include those addressed in the following paragraphs.

Wearing Characteristics.   The location and use of a resilient floor covering should be taken into consideration when deciding which flooring material to select. Using materials with a no-wax coating in a commercial application where they are subject to much traffic will result in a rapid deterioration of surface appearance, for example.

Maintenance Characteristics.   The level of maintenance to be expected should be ascertained and used as a determinant in deciding which material to select.

Substrates.   The type of resilient flooring material that should be used is influenced by the material and location of the substrate. The resilient flooring manufacturer's recommendations should be followed. In general, the following rules apply.

- Asphalt and vinyl composition tile may be applied on concrete slabs that are in contact with earth where no waterproofing is present, and on suspended slabs at or below grade. A vapor retarder does not classify as waterproofing for this purpose. Vinyl and rubber tile, and sheet vinyl products should not be used under these conditions.
- Sheet vinyl with a moisture-resistant backing and vinyl, vinyl composition, rubber, and asphalt tile may be applied on slabs that are in contact with earth and waterproofed with a membrane and for suspended slabs at or below grade that are waterproofed with a membrane. A vapor retarder does not classify as waterproofing for this purpose.
- Vinyl, vinyl composition, rubber, asphalt, and cork tile; sheet vinyl and rubber; and linoleum may be installed on suspended floors above grade with ventilated areas beneath the slab.
- Vinyl and rubber tile with special adhesives, asphalt tile, vinyl composition tile, and sheet vinyl with special adhesive may be installed over slabs containing radiant heating coils. Rubber or vinyl tile or sheet vinyl may not be used in slabs having radiant heating coils, however, when the slab is in contact with the earth or suspended at or below grade, unless a waterproofing membrane is in place.
- Vinyl, vinyl composition, rubber, and asphalt tile, and sheet vinyl should not be applied directly onto lightweight concrete having a density of less than 90 pounds per cubic foot. Such lightweight concrete

tends to be too weak to support the bond of the resilient flooring adhesives. Instead a normal-weight concrete topping should first be applied to receive the resilient materials.

- Resilient flooring should not be installed over a wood floor constructed over a crawl space that is not properly ventilated.
- Resilient flooring should not be installed over a wood floor installed on sleepers fastened directly to a concrete slab that is on or below grade.

Refer to "Preparation of Concrete Substrates" later in this chapter for methods of determining the suitability of concrete to receive resilient flooring, based on moisture content of the concrete.

Material Size. Resilient flooring materials for new projects may be selected because of their size. Tile size is more likely to be an aesthetic consideration, but sheet size for sheet materials may also be made for sanitary reasons. Sheet materials are available in rolls that vary in width between 4 and 12 feet. Larger sheets may be selected so that there can be no joints in some spaces, such as corridors and small rooms.

Material Static Load Limit. Resilient tile is seldom selected based on its static load limit. Many manufacturers, in fact, do not list that characteristic in their product literature. Conformance with Federal Specification SS-T-312 establishes the salient characteristic relative to load bearing, and most resilient tile products at least claim to comply with SS-T-312.

The static load limit is often considered when resilient sheet materials are being selected, however, especially where heavy loads are to be encountered.

Chemical Resistance. Vinyl sheet materials are often used where chemical resistance is an important consideration. Consequently, many vinyl sheet flooring products have been tested for chemical resistance, and their manufacturers regularly publish the results.

Even some common materials can cause adverse reactions in some resilient materials. Both asphalt and rubber flooring, for example, are susceptible to damage by oils and fats and should, therefore, not be used in commercial kitchens or other locations where they are likely to come into contact with such materials.

Cost. Too often, resilient flooring is selected based on initial cost alone. While important, initial costs should be weighed against the other factors listed above and the overall costs, which includes the cost of maintenance and a factor for length of service.

There are no later problems that are associated specifically with material cost. Low cost often indicates lesser quality, however, and so may predict future problems.

Tile

Resilient tile materials include vinyl, vinyl composition, rubber, asphalt, and cork tiles. Resilient tile is used where cleanliness and sanitary control is not so critical. Tile is also used where the substrate precludes use of sheet materials.

Tile Standards.   Resilient tile is classified in accordance with U.S. GSA Federal Specification SS-T-312 "Tile, Floor: Asphalt, Rubber, Vinyl, Vinyl Composition" or U.S. GSA Federal Specification L-F-475, "Floor Covering, Vinyl, Surface (Tile and Roll), with Backing."

Vinyl Tile.   Although it is slightly less resilient than rubber tile, homogeneous vinyl tile is probably the best resilient tile available for all-around performance in most uses. Its only major disadvantage is that it has a low resistance to cigarette burns. Vinyl tile is a blended composition of a binder, fillers, and pigments, which is stabilized against heat and light deterioration. The binder, which is composed of one or more vinyl resins and plasticizers, should be at least 34 percent of the tile by weight. The vinyl resin, which should be not less than 60 percent of the binder by weight, should be polyvinyl chloride or a vinyl chloride copolymer, of which not less than 85 percent is vinyl chloride. A thin protective coating may be applied to vinyl tile. Vinyl tile should be type III, in accordance with Federal Specification SS-T-312. It should be accurately cut, square, and smooth faced.

Many colors are available, including plain, marbleized, and mottled colors. Most extend through the entire thickness of the tile. Many patterns are also available, including flat tile and simulated wood, ceramic tile, quarry tile, terrazzo, brick, slate, and marble.

Vinyl tile is available in many sizes, including 6 by 6, 6 by 12, 9 by 9, 12 by 12, 12 by 18, 18 by 18, and 36 by 36 inches. Some manufacturers also have 36-3/4 by 36-3/4 inches untrimmed vinyl tile slabs for field trimming. Small inserts, such as those 3 by 3 inches, are available for use as accent pieces. Various manufacturers list their vinyl tile's thickness as 1/16, 1/8, 0.080, or 0.10 inch. Some special units are also available. One example is a vinyl unit with a simulated strip wood finish that comes either 3 by 36 or 9 by 36 inches. Both sizes are 0.10 inch thick.

Solid homogeneous vinyl tile is also available in conductive and static dissipative versions, for use in hospital surgical and anesthetizing spaces, and in computer rooms and chemical laboratories, or wherever explosive elements are used or static electricity presents a hazard. Static dissipative tile is used where a higher level of protection is necessary than that which can be afforded by conductive tile. Both types are available 1/8 inch thick, and in several sizes, including 12 by 12, 24 by 24, and 36 by 36 inches.

Vinyl Composition Tile. Up until a few years ago, before the dangers associated with asbestos became well known, vinyl composition tile was composed of a thermosetting binder, asbestos fibers, mineral fillers, and pigments. The binder consisted of polyvinyl chloride resins, or a copolymer resin compound, with suitable plasticizers and stabilizers. Modern vinyl composition tile contains essentially the same elements, except that the reinforcing fibers are not asbestos. Vinyl composition tile should be type IV, in accordance with Federal Specification SS-T-312, and either composition 1 (non-asbestos) or composition 2 (with asbestos). It should be accurately cut and square.

Vinyl composition tile is available 9 and 12 inches square and 1/8 or 3/32 inch thick. Tiles 0.080 or 1/16 inch thick are also available, but are mostly used in residences.

Many colors are available, including plain colors, and marbleized and mottled colors. The color of the wearing layer of plain color tile should run through the full thickness of the wear layer, but may or may not run through the entire tile thickness. Marbleized and mottled patterns usually run through the full thickness of the tile.

The face of vinyl composition tile may be either smooth or textured (embossed). In general, the depressed areas of embossed tile should be not more than one-third of the total surface area. Many patterns are available, including those that simulate marble, slate, tile, and brick.

Rubber Tile. There are several kinds of rubber tile in use today. One type is made from rubber tile components. That type is beyond the scope of this book. The type of rubber tiles addressed here are the kind most used in commercial buildings today. They are homogeneous rubber. The rubber is made from first quality 100 percent natural or synthetic virgin rubber and pigments, fillers, stabilizers, integral waxes, and soil-releasing agents. The rubber material should be odorless and stain resistant. Some rubber tiles conform with the requirements stipulated in Federal Specification SS-T-312, others do not.

Some rubber tiles are flat. Others have a raised profile (Fig. 4-1). Rubber tiles are also available with raised patterns on the backs as well as on the wearing surface. Available raised wearing surface profiles include:

Low-Profile Raised Discs. Discs may be low-profile types, which are between 1 inch and 1.18 inches in diameter, and range in height from

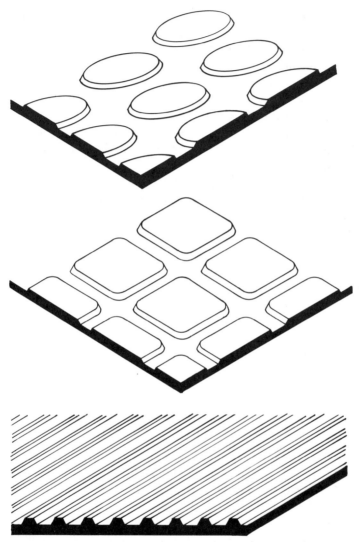

Figure 4-1   Typical raised profile configurations for rubber tile and sheet materials.

0.020 to 0.027 inches. Total thickness at the discs should be between 0.150 and 0.160 inch.

Medium-Profile Raised Discs. Discs may be medium-profile types, whose discs range in diameter from 1 inch to 1.024 inches and are 0.039 inch high. Overall thickness at the discs is 0.172 inch.

High-Profile Raised Discs. Two types are available. One has discs that are 1 inch in diameter and between 0.050 and 0.059 inch high. Overall thickness at the discs is 0.175 inch. The second type has discs with a smaller diameter (0.86 inch or less) and a height not less than 0.078 inch. The overall thickness is at least 0.158 inch.

Low-Profile Raised Squares. The squares are between 1.26 and 1.94 inches on a side and between 0.020 and 0.024 inch high. Overall thickness at the raised portion is 0.110 inch.

Medium-Profile Raised Squares. The squares are between 1.25 and 1.65 inches on a side and between 0.031 and 0.039 inch high. Overall thickness is 0.156 inch.

Other Profiles. Some manufacturers offer tiles with other patterns. Some materials intended for use on stair landings, for example, have a diamond-shaped surface pattern. Others have discs much smaller than those mentioned earlier, raised H-shapes, parallel ribs (Fig. 4-1), and other shapes.

Rubber tile is available in a variety of sizes. Most are squares with one of the following side lengths: 12, 17-13/16, 18, 18-1/8, 19.625, 19.69, 19.8, 24, 36, 39.37, and 36-3/4 inches. Manufacturers list many thicknesses, including 3/32, 1/8, 3/16, 0.10, 0.130, 0.137 inches, and 3.0, 3.5, 4.5, and 5.0 millimeters.

Many colors and patterns are available. Most extend completely through the tile. Colors are plain, mottled, marbleized, or simulations of other materials.

Asphalt Tile.   An early form of resilient tile, asphalt tile may be found in many older structures. Early versions contained asphalt or other bituminous binders, asbestos fibers, mineral fillers, and pigments. Asphalt tile is still available today from at least one manufacturer, but it is made with thermoplastic binders, and non-asbestos fibers. Federal Specification SS-T-312 classifies asphalt tile as type I, but the manufacturer of the currently made product classifies it as type IV (vinyl composition), composition 1 (asbestos-free).

Asphalt tile should be accurately cut and square, with a smooth face. It is available today in 9 by 9 by 1/8 inch tile. Other sizes may occur in older installations. About nine colors are available, but the only pattern is marbleized.

Cork Tile.   Resilient tiles made from cork are also available. Standard cork tile, which may be unfinished for field-finishing, usually with urethane, or factory-finished with a high-gloss polyurethane coating, is usually 12 by 12 inches. Thicknesses are 3/16 and 5/16 inches.

Cork tile is also available with a 0.02-inch thick vinyl wear layer, 0.09

inch cork, and a 0.016-inch thick PVC foil backing. This material is available in tiles that are either 12 by 12, 6 by 6, or 6 by 36 inches in size. The total thickness is 0.126 inch.

There is also available a cork flooring material that is finished with a natural-finished wood veneer wear layer. It is available in pieces that are 12 by 12, 3 by 36, or 6 by 36 inches in size. Their thickness is 0.126 inch.

Sheet Flooring

Resilient sheet materials include vinyl, rubber, and linoleum. Resilient sheet material is used where cleanliness and sanitary control are important.

Standards. Resilient sheet flooring is classified in accordance with Resilient Floor Covering Institute (RFCI) *SV-1, Resilient Floor Covering Institute Recommended Specification for Resilient Floor Covering—Vinyl Plastic Sheet* and U.S. GSA Federal Specifications L-F-475, "Floor Covering, Vinyl, Surface (Tile and Roll), with Backing" and L-F-001641, "Floor Covering, Translucent or Transparent Vinyl Surface with Backing."

Vinyl Sheet Flooring. There are two universal characteristics of vinyl sheet flooring. First, the surface is continuous, smooth, washable, easily decontaminated, chemical resistant, and abrasion resistant to foot traffic. Second, a major portion of the flooring is polyvinyl chloride. Beyond that, vinyl sheet flooring products vary tremendously. For example, some of the many different combinations of materials that make up the various types of vinyl sheet flooring include:

Filled vinyl with a fibrous backing
Filled vinyl with a foam backing
Filled vinyl without a backing
Unfilled vinyl with a foam interlayer and backing
Unfilled vinyl with a vinyl interlayer and backing
Unfilled vinyl with a reinforced vinyl backing
Unfilled vinyl with a reinforced vinyl interlayer and a foam backing
Raised profile vinyl with no backing

Most older vinyl sheet flooring products can be classified according to one or more of the standards referenced under the previous heading. Some newer products, however, are difficult to classify. Each of the three available standards is usable for only a portion of the available products. The following description is not intended to repeat the requirements in those documents, but rather to acquaint the reader with the documents and their contents.

Anyone with a problem related to vinyl sheet flooring should obtain the three referenced publications.

RFCI *SV-1* contains requirements only for sheet vinyl products that have a backing. It divides those products into two use-categories. Category I includes products that may be applied on, above, or below grade. Category II includes products that may be applied only when the floor has 16 inches of cross-ventilated space beneath it. *SV-1* also classifies wearlayers by type and material, backing by groups indicating in which use-category each can be used, and grade designations which designate wearlayer and overall thickness.

Federal Specification L-F-475 is applicable to both tile and roll products with filled vinyl wearlayers and either organic or inorganic backings.

Federal Specification L-F-001641 is applicable to vinyl sheet flooring with transparent or translucent wearing surfaces. It classifies such materials by backing type and thickness and lists intended uses for the three types that it designates.

Vinyl sheet flooring is available in many colors and patterns. It may be a solid color, or be marbleized, mottled, or have several colors of chips embedded in a transparent wearlayer. It may have a printed pattern of lines or grids. Some vinyl sheet products have a texture that imparts nonslip properties to it. Some have embedded abrasive grains, such as aluminum oxide.

Vinyl sheet flooring is available in rolls ranging from 4 to 12 feet in width. Available overall thicknesses of flat sheets vary from 0.060 to .197 inch, depending on the type of interlayer and backing and the manufacturer. The overall thickness of raised pattern sheets varies between 0.100 and 0.106 inch. The wearlayer thicknesses of flat products vary from 0.020 to 0.080 inch. Some raised pattern materials have wearlayers that are 0.035 inch thick. In solid materials without a backing, of course, the wearlayer is the full thickness of the sheet.

Most vinyl sheet flooring requires waxing to bring out its full beauty, and to make it less easily damaged during use. Some vinyl sheet flooring, however, has a factory finish that does not require waxing. There are two kinds of finish used on such "no-wax" flooring. One type is intended for residential use only. It achieves the no-wax feature by means of an applied coating, which is usually urethane. The second type relies on the physical properties of the wearlayer material to achieve the finish. This type is usable in commercial installations. Neither finish is permanent, however, but both can be repaired.

The joints in vinyl sheet flooring may be simply butted. Where a continuous floor is required, joints may be welded. Vinyl flooring that does not have a backing may be heat welded using a special heating tool and vinyl welding rods. Joints in other vinyl sheet flooring may be welded using chemicals.

Because of the difficulty of classifying some vinyl sheet flooring, a particular product is often selected as the only acceptable product or as a standard against which other possible products will be judged. Product selection is usually based on the selecter's personal preferences, combined with other factors, such as the location within the building, whether the product under consideration is available in the desired sheet width, whether the product's joints can be welded and by which means (heat or chemical), the product's chemical resistance and static load limit, and the available colors and patterns.

Where generic selection is required or desired, the three standards discussed may be used, but such use may eliminate some products. The standards must be used very carefully, however. It is easily possible to combine requirements from the standards in such a way that no product manufactured can meet all of them.

Sheet Rubber Flooring.   There is little difference between the rubber tile discussed earlier and sheet rubber flooring except, of course, that the latter comes in rolls. Some sheet rubber flooring manufacturers say that their product complies with the requirements of Federal Specifications SS-T-312, which may be generally true, but SS-T-312 actually applies only to tile. Others say that their product complies with the requirements of Federal Specification ZZ-F-461, which also may be true, but ZZ-F-461 was withdrawn some years ago. In actuality, there is no current industry or government standard for the performance of rubber sheet flooring.

Good sheet rubber flooring is made from fully homogeneous natural or synthetic rubber. Many colors and patterns are available. Most extend completely through the material. Colors are plain, mottled, or marbleized. It is available in several thicknesses, including 3/32, 1/8, and 3/16 inch, and in nominal 36-inch wide rolls.

Sheet rubber flooring is available either flat or with a raised pattern of the same types discussed earlier for rubber tile (see Fig. 4-1) or another pattern. Some flooring is made specifically for use on stair landings, and comes in patterns that match those used on stair treads, which are discussed later in this chapter.

Linoleum.   The original sheet flooring, linoleum was a set mixture of ground cork and linseed oil bonded to a burlap backing. Modern linoleum includes cork and linseed oil as well as wood dust, pigments, and natural resins. Its backing is usually jute.

Linoleum is available in many colors, several thicknesses and weights, and in sheet widths of about 6 feet.

Miscellaneous Resilient Units and Accessories

Complete resilient flooring applications require many types of miscellaneous resilient units and accessories. Among them are resilient bases; stair risers, stringers, and treads; cove caps; reducer strips; carpet edge guards; thresholds; and many others.

Standards.   Resilient wall base and stair stringer materials are classified in accordance with Federal Specification SS-W-40, "Wall Base: Rubber and Vinyl Plastic."

Stair treads are classified in accordance with Federal RR-T-650, "Treads, Metallic and Non-Metallic."

Resilient Wall Base, Stair Riser, and Stringer Material.   Base and stringer materials may be either rubber complying with Federal Specification SS-W-40 Type I, or vinyl complying with Federal Specification SS-W-40 Type II.

Where the flooring is resilient flooring or another hard surface material, bases are usually standard or butt-type set-on premolded cove-style (Fig. 4-2). Where the flooring is carpet, bases are usually standard premolded

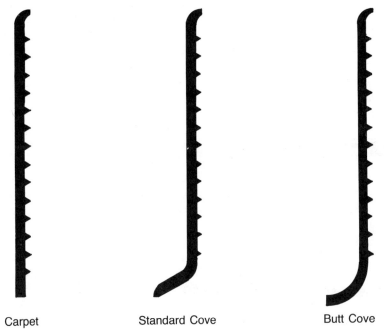

Carpet   Standard Cove   Butt Cove

Figure 4-2   Resilient base types.

carpet-base (without a cove) (see Fig. 4-2). Bases may be 2-1/2, 4, or 6 inches high and either 1/8 or 0.080 inch thick. Corners in a base may be premolded or made in the field from sections of base. Bases are available in as many as 14 plain colors. They come in short rolls (usually about 4 feet long) and in long rolls of usually about 100 or 120 feet.

The base for sheet flooring, especially where the joints in the flooring are welded, are often made by turning the flooring material up the wall. Such bases are often terminated in a cap of some sort. Some such caps are resilient (Fig. 4-3). Others are metal, such as aluminum. Refer to "Installation" for additional details.

Stair risers are available in rubber and vinyl and in colors to match the various treads available. Risers are usually 6 or 7 inches high, 0.08 or 0.10 inch thick, and up to 6 feet long. Wider riser materials are also available for use where higher stair risers occur. Risers are also available in 50-foot

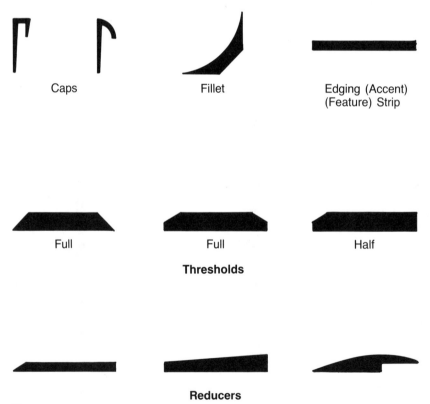

Figure 4-3 Accessories for use with resilient flooring.

long rolls. The bottom of the risers is often coved similar to wall bases, but flat risers without a cove are also available.

Stringer material comes in both rubber or vinyl to match the base or stair treads used. It is available in sheets up to 6 feet long and 50-foot long rolls that are 10, 12, or 36 inches wide and 0.080, 0.085, 0.10, or .125 (1/8) inch thick.

Stair Treads and Nosings.   There are many forms of stair treads and nosings available. Most comply with the requirements of Federal Specifications RR-T-650, ZZ-T-001237, or both. Treads and nosings may have a square or round shape (Fig. 4-4). Treads may be smooth surfaced or have a surface

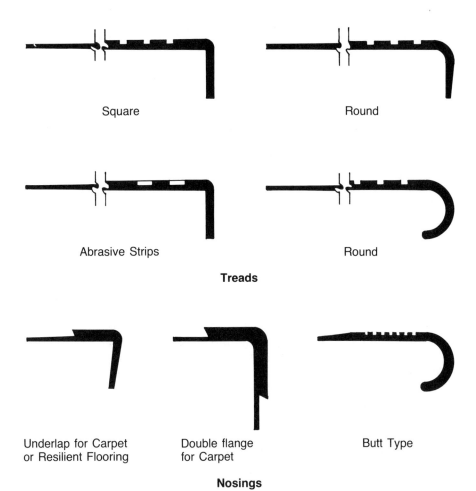

Square   Round

Abrasive Strips   Round

Treads

Underlap for Carpet   Double flange   Butt Type
or Resilient Flooring   for Carpet

Nosings

Figure 4-4   Resilient stair treads and nosings.

pattern. Many patterns are available, including diamonds, discs, rectangles, squares, and parallel ribs. Treads are also available with embedded abrasive strips. Many colors are available in plain, mottled, or marbleized patterns.

Treads are available to fit many tread widths. Others can be cut to fit from wider material. Most treads taper in thickness from nose to back. Thicknesses vary from 1/8 to 1/4 inch at the nose and from 1/4 to 3/32 at the back.

Nosings have shapes similar to that of treads, but are designed to fit beneath the tile or carpet at the back or butt with the finish flooring (see Fig. 4-4).

Resilient Edging Strips.   Edging strips (also called accent or feature strips) are strips of rubber or vinyl. They are usually 1/2 or 1 inch wide, 36 inches long, and the thickness of the resilient flooring (see Fig. 4-3).

Resilient Reducer Strips.   Reducer strips (see Fig. 4-3) are strips of rubber or vinyl that are tapered to make a smooth transition from one flooring material to another. They may be the thickness of the resilient flooring on one edge and taper to zero on the other for use where resilient flooring terminates at a concrete subfloor. They may be the thickness of the resilient flooring on one side and taper up to the thickness of installed ceramic tile on the other side where resilient flooring abuts ceramic tile and there is no other threshold.

Reducer strips range in width from 1 inch to 1-1/2 inches wide and are usually 36 inches long.

Welding Rod for Vinyl Sheet Flooring.   Welding rods should be the flooring manufacturer's standard PVC bead and may be clear or colored as selected from the manufacturer's standard colors.

Miscellaneous Accessories.   Some miscellaneous accessories are shown in Figure 4-3 and 4-5. There are many others available. They are used for carpet, ceramic tile, and resilient flooring. There are caps for turned-up bases made from carpet or resilient materials. There are many types of carpet dividers and thresholds. There are thresholds as well for resilient and other flooring materials. There are fillets used to back up resilient bases made from the same material as the floor. There are corner guards, tub moldings, and finish moldings. There is a resilient accessory made for virtually every condition where resilient flooring, carpet, or ceramic tile terminates or meets another material. Both vinyl and rubber accessories are available in colors to match or contrast with most resilient materials.

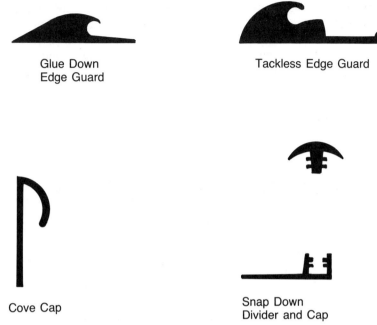

Glue Down
Edge Guard

Tackless Edge Guard

Cove Cap

Snap Down
Divider and Cap

Figure 4-5   Resilient accessories for use with carpets.

Underlayment, Primers, and Adhesives

Underlayments include mastic and other liquid-applied types and board types. Liquid-applied type underlayments and primers and adhesives should be waterproof types recommended by the manufacturers of the resilient material. They may vary depending on the resilient material, substrate, location, and condition. Asphalt emulsions or other nonwaterproof types of mastic underlayments, primers, and adhesives should not be used.

Underlayments.   Both liquid-applied and board type underlayments should be made for the purpose and compatible with both the substrates and the resilient material adhesives.

Latex Underlayments.   Most mastic underlayments contain latex or polyvinyl acetate as a binder and Portland cement. They should be easily trowelable. It should be possible to apply them to a feather edge. These products are usually manufactured by, or are available from, the resilient flooring manufacturer.

Self-leveling Cementitious Underlayment. There is available today a family of proprietary cement-based products designed for use as leveling toppings and underlayment on a variety of floor substrates, including wood, plywood, concrete, terrazzo, and ceramic tile. They are not universally applicable, however, so their usability in a particular situation should be verified. One product might be usable, for example, over adhesive residue, while another is not. One may work on slabs-on-ground. Another may not. These products are sometimes used in new construction, but their primary use is to level existing substrates to receive new flooring.

Board-type Underlayment. The best board-type underlayment is plywood. For normal installations the plywood underlayment should have been rated by the American Plywood Association as an APA Underlayment grade. Where the flooring will be subjected to unusual moisture, an APA C-D Plugged Exposure 1 grade should be used. For more information about plywood underlayment selection, refer to APA Data File "Installation and Preparation of Plywood Underlayment for Thin Resilient (Non-Textile) Flooring."

Particleboard, waferboard, chipboard, lauan plywood, and tempered or untempered hardboards are not recommended for use as underlayment beneath resilient flooring by some authorities. Before such a material is used, its manufacturer's warranty for the purpose should be ensured and the resilient material manufacturer should concur with the selection.

Primers. The type and brand of primer recommended by the resilient material manufacturer should be used.

Adhesives. The type and brand of adhesive recommended by the resilient material manufacturers for each resilient material and condition should be used. Where a bond is more difficult because of the existence of foreign materials or cleaning cannot be done completely, the manufacturer may recommend a special adhesive. Solvent-based adhesives, for example, may be substituted for water-emulsion or water-dispersion adhesives in some cases. Different manufacturers may recommend other types, but the following are some typical adhesive types often used:

- For vinyl composition tile: Asphalt cut-back type.
- For bases, risers, and stringers: Specially formulated waterproof high-strength resin-rubber type.
- For vinyl sheet flooring: Installation adhesive is often a rubber-based, stabilized type. Seam adhesive is often, but not always, epoxy-based.
- For rubber sheet flooring: Latex-based type.

- For rubber tile: Trowelable or chemical-set type.
- For stair treads and accessories: Resin-based, alcohol-soluble type.

Cleaning and Polishing Materials

Cleaning Materials. Cleaners vary somewhat, depending on their use. Cleaners intended for use on new resilient floors are neutral chemical cleaners that are free from damaging alkalis and acids, and free from oils and abrasives. Some are UL classified as to slip resistance.

Cleaners intended for use on existing resilient floors are fast-acting types formulated to clean tough dirt and grime. They should be pH safe for resilient floors, and should leave no dulling soap scum build-up.

Sealer. Materials specifically formulated to form a protective seal on resilient materials and an undercoater for the finish on such materials is called a sealer. Some sealers are UL classified as to slip resistance.

Finish. Several types of polish are routinely used on resilient flooring. On some types of flooring a urethane finish will eliminate the need for waxing. An excellent wax finish can be had by using a self-annealing, self-polishing, water-emulsion wax, made of 100 percent prime No. 1 carnauba wax. Most modern finishes, however, are based on acrylic polymers with a small amount of wax added. Some finishes are UL classified as to slip resistance.

Paste wax, paint, varnish, lacquer, shellac, and other solvent-containing coatings should not be used on resilient products containing rubber.

The actual finish selected should be the one recommended by the resilient material manufacturer.

Resilient Flooring Installation

Requirements Common to All Resilient Flooring

Resilient flooring is used in residential, commercial, institutional, and industrial buildings on all sorts of floors, including elevated-access and elevator floors. It is usually installed over the entire exposed-to-view floor area in spaces where it is used, from wall to wall and into toe spaces, door revels, and floor level openings. It is also often installed in closets that adjoin spaces where it is used. Normally, however, fixed casework and cabinetwork with closed bases, floor-mounted heating and air conditioning unit enclosures, and fixed equipment items are installed before resilient floors are laid. Occasionally, though, especially in laboratories and related spaces, and in other spaces where such items are likely to be changed before the flooring wears out, such items are installed on top of the resilient flooring.

A resilient base is often used where resilient flooring or carpet occur. It is usually applied to all wall surfaces and often to the toe base of millwork, cabinets, casework, lockers, and equipment where a finished base is not an integral part of those units. It may also be applied in other locations. Loose bases are usually used with tile and sheet materials with unwelded joints, and may also be used with sheet materials with welded joints. Sheet flooring with welded joints are sometimes turned up adjacent wall surfaces to form an integral base.

Substrates should be installed, prepared, and finished strictly in accordance with the instructions of the resilient materials manufacturer and standard industry practices.

Before resilient flooring installation begins, a working layout should be prepared for each area that is to receive resilient flooring, especially sheet flooring. It should show the location of each seam; the location and types of bases, edgings, and accessories; changes in color; and other pertinent installation details.

It is customary to leave at the site, usually stored in factory cartons or rolls, extra material of the same type and colors as was used on the job for the owner's use in making repairs. The amount of flooring left is normally about three percent of the amount used in the building, divided in approximate proportion to the type of flooring and the number of colors or patterns used. The minimum is usually not less than one carton of each tile or one roll of each sheet flooring. The minimums should, of course, be increased for larger projects. Extra resilient base material, stair treads, and stair stringer material should also be left to make repairs in those items. Sufficient adhesives, accessories, joint materials such as welding rods, underlayment, and other related items should be left to install the stored resilient materials and to make incidental repairs to those items.

Resilient flooring materials should be delivered to the job site in the manufacturer's original, unopened containers, with the manufacturer's label intact. Materials should be packed, stored, and handled carefully to prevent damage.

Materials should not be permitted to freeze. Materials and spaces should be maintained at not less than 65 degrees Fahrenheit for 48 hours before and until 48 hours after installation, and above 55 degrees Fahrenheit after that. The resilient material should be stored in the space where it will be installed for at least 48 hours before it is installed.

Adequate ventilation should be provided in the space where resilient materials are being installed. The resilient flooring manufacturer's recommended safety precautions should be followed.

Flooring materials should be cut and scribed to fit neatly at walls, breaks, recesses, pipes, door frames, cabinets, fixtures, equipment, thresholds, and edgings. Flooring should be butted tightly to other materials and surfaces.

When installing resilient materials, only proper adhesives, without adulteration or reduction, should be used. Use should be in accordance with the manufacturer's instructions. Only that area that can be covered with resilient material within the manufacturer's recommended working time of the adhesive should be covered with adhesive.

Work should not be started in a space until the work of other trades, including painting, has been substantially completed and, during winter months, not until the permanent heating system is operating. The moisture content of concrete, building air temperature, and relative humidity should be within the limits recommended by the flooring manufacturer as determined by moisture tests.

It should be verified that surfaces to receive resilient materials are in proper condition to begin work. Installation should not proceed until unsatisfactory conditions have been properly corrected.

Preparation of Substrates.   When preparing substrates to receive resilient materials, the printed recommendations of the resilient materials manufacturer should be followed, unless they are obviously incorrect.

Since resilient flooring is thin and flexible, it tends to show even small imperfections in the substrates. Substrates to receive resilient flooring should be smooth, level, and in plane to within 1/8 inch in 10 feet, when measured with a 10 foot straightedge. Imperfections and foreign materials that would telegraph through the flooring should be removed. Substrates should be cleaned thoroughly to completely remove all traces of oil, wax, dust, paint, and other materials that would reduce the adhesive bond or harm the flooring.

Preparation of Wood Substrates.   Installation of resilient flooring over new strip wood flooring in a new building is rare. Such an installation might be found in an existing building, however. Refer to "Installing Resilient Flooring over Existing Materials" later in this chapter for recommendations for the preparation of wood strip flooring to receive resilient flooring.

Preparation of Plywood Substrates and Underlayment.   Underlayment should be provided and the substrates primed if so recommended by the resilient materials manufacturer. Board underlayment should be properly installed and structurally sound, smooth, clean, and ready to receive resilient materials. Nails should be set. Joints should be sanded smooth. Gouges, chipped areas, and open joints should be filled with a hard-setting filler formulated for the purpose. Joints should be left slightly open to permit expansion. Dried filler should be sanded smooth.

Preparation of Concrete Substrates. Concrete substrates should be screeded and troweled to a smooth plane surface free of score marks, grooves, depressions, and trowel marks. Ridges and other defects which would telegraph through the flooring should be ground smooth. Membrane-forming curing compounds should not be used. Where they are used, they must be completely removed before resilient flooring is installed. Removal may require sanding (which is seldom completely successful), sandblasting, grinding, or polishing off with a wire brush in a polishing machine. Concrete surfaces should be scraped to remove dirt and foreign matter and brushed clean. Uneven places should be leveled by chipping, filling, or grinding. Cracks 1/16 inch wide or wider should be filled with a suitable crack filler. Remaining minor construction joints, grooves, cracks, holes, and rough areas should be filled with a latex underlayment to provide uniform surfaces. Latex underlayment should also be used to smooth and level the slab. Loose particles should be removed. Chalky, dusty surfaces should be vacuumed clean. Alkaline salts should be completely removed from the slab surface. The cleaned slab should be primed when recommended by the flooring manufacturer.

Concrete to receive resilient flooring must be dry enough to permit the adhesives to bond properly and to prevent damage to the flooring. Alkaline salts exuding from curing concrete can damage resilient flooring. Water in slabs can cause adhesives to fail. The limitation on the amount of water present is dictated by the various resilient flooring products manufacturers. Some linoleum manufacturers, for example, recommend a maximum moisture content of 3.5 percent. The Rubber Manufacturers Association recommends that not more than 3 pounds of water vapor per 1000 square feet egress from a floor to receive rubber flooring products in a 24 hour period. The same rule applies for solid sheet vinyl. The Portland Cement Association says that it will take a concrete slab at least 90 days to reach that level. Under some conditions, it will take longer. Some resilient material manufacturers recommend time limits or specific test results.

Restrictions on the types of resilient flooring which may be installed over various concrete substrates is listed earlier in this chapter under "Characteristics Common to All Types of Resilient Flooring." Each of those substrates should be tested to ascertain its moisture content before resilient flooring is installed. The types of concrete substrates classify as either slabs placed in contact with earth or supported slabs.

Slabs that are placed on earth are never really dry unless a capillary water barrier and a membrane waterproofing layer are installed beneath them. Therefore, some resilient flooring manufacturers limit to asphalt and vinyl composition tile the products which may be used on slabs placed on earth when a waterproofing membrane is not present. Even those materials should not be installed until at least three months after the concrete is placed and, even then, not until the slab is tested and proven to be dry.

Other slabs should be permitted to dry about six months and then should be tested for moisture content before resilient flooring is installed over them.

The actual curing time for concrete slabs may vary, depending on the type of curing compound used; whether a breaking compound was used, such as is used in tilt-up construction; whether an antidusting compound was used; the amount of troweling; and the type of aggregate used. More troweling may mean longer setting time. Using cinder, pumice, and other lightweight aggregates may cause concrete to cure slower.

There are several methods for testing concrete slabs to determine if they are sufficiently dry to receive resilient flooring. They include, but are not necessarily limited to, the following:

- Electrical moisture indicator test: An electrical moisture indicator works by measuring the resistance of electricity through the water in a concrete slab. To use this method, it is necessary to drill holes in the slab to receive pins. This method is used for slabs in contact with earth.

- Primer or adhesive strip test: After the slab is apparently dry, and has been allowed to cure for two months if on earth, and six months if supported, several small patches about 24 inches square of primer or adhesive should be placed on the slab in question. If after the primer or adhesive has been down 24 hours it bonds securely to the slab, the resilient material may be installed. If the primer or adhesive can be peeled from the floor using a putty knife, the slab must dry more. This method may be used on slabs placed on earth or supported slabs, but only when asphalt or vinyl composition tile will be used.

- Polyethylene sheet test. A simple test to determine the suitability of a concrete floor for resilient flooring application involves taping a 48-inch-square piece of polyethylene or other plastic film to the concrete. If moisture accumulates on the plastic film within 24 hours, the concrete is too wet to successfully receive resilient flooring.

- Mat test: A mat test is executed by taping polyethylene sheeting, linoleum, or sheet vinyl flooring pieces at least 24 inches square to the concrete surface over a band of water-soluble adhesive and a band of water-resilient adhesive. After 72 hours, an inspection is made to determine the condition of the adhesives. If the water-soluble adhesive is partly or completely dissolved, or the water-resistant adhesive is stringy or has little bond, there is too much water in the slab. If the test is unsatisfactory, the concrete must be permitted to dry further or a more water-resistant flooring used. This method may be used for testing slabs in contact with earth and supported slabs. It may also be used for testing other types of substrates. It is particularly useful when the flooring is vinyl sheet or rubber or solid-vinyl tile, or when paint, oil, or curing compound has been removed, regardless of the flooring material to be used.

- Humidifier test: When the flooring is moisture-sensitive, as are such flooring products as linoleum, cork, and rag-felt-backed sheet vinyl, a relative humidity test is often conducted. In such a test, a meter is placed on the slab and sealed in place beneath a layer of polyethylene sheeting. The polyethylene sheet is sealed to the floor using tape. The length of test depends on the slab thickness, with a minimum of 24 hours and a maximum of usually 72 hours. The meter should read less than 80 percent before resilient flooring installation is attempted.

- Chemical test: Granulated anhydrous calcium chloride is placed in a dish and sealed beneath a glass cover. Holes are drilled in the slab, also beneath the glass cover. The test is continued for 72 hours. If beads of moisture appear on the cover glass or the anhydrous calcium chloride starts to dissolve, there is too much moisture in the slab. This test is best for testing suspended concrete slabs.

- Actual-material test: When concrete has been coated with a curing agent, or another material has had to be removed using sanding, grinding, sandblasting, wire brushing, or another similar method, several pieces of the actual resilient material to be installed may be installed as a test. The test pieces should be left in place from three days to two weeks, depending on the products used. If the bond is still strong, the remaining resilient flooring may be installed. Some manufacturers suggest specific methods to be used when performing actual-material tests, including materials to use, procedures, and necessary elapsed times, and should be consulted when such testing is contemplated.

Adjoining surfaces to receive resilient materials should be in the same plane. Adjacent surfaces to receive different finish floor materials should be in proper planes to permit acceptable transitions.

Surfaces to receive resilient materials should be clean and dry before resilient material application is started. Such surfaces should be swept and vacuum cleaned immediately before resilient materials are installed.

Installing Resilient Tiles

Tile should be laid smoothly, without air pockets, in accurate alignment, with joints brought in tight contact, and without raising or puckering at the joints, telegraphing of adhesive spreader marks through the tile, or other surface imperfections. Tile should be tightly cemented to the substrate.

Field tile units may be laid in checkerboard fashion, with the grain reversed in alternate units, or with the grain in all tile running in same direction.

Border tile and feature tile may be used.

Most tile is laid starting in the center of rooms or spaces and using full size tile for the field and borders where borders are used. Slight adjustments may be made to permit the tile alignment to continue through door openings where resilient tile occurs in adjacent rooms. Tile should be laid to avoid the use of cut widths less than 4 inches wide against walls. Edge units should be approximately the same width on each side of the room, when possible.

Much resilient tile is installed with the edges parallel with the room or space walls. Other arrangements are also used, of course. A common exception involves laying the tiles with their edges at 45 degrees to the walls.

It is normal to match the tiles for color and pattern by using them from cartons in the same sequence as they were manufactured and packaged.

Latex underlayment should be used as necessary to ensure that the tile finishes in the proper plane relative to adjacent floor finishes, and so that edging strips and reducers finish flush with the top of the resilient flooring.

Broken, cracked, chipped, or deformed tile should not be used.

Installing Resilient Sheet Flooring

Sheet flooring may be installed with butted joints or with heat-welded or adhesive sealed joints. The wall base at resilient sheet flooring may be separate or integral.

Sheet flooring that is to have the joints heat-welded or adhesive sealed should be installed by workers who are certified by the flooring manufacturer as qualified to make such seams on the particular product being installed.

The following requirements are applicable to most forms of sheet flooring. They may not be fully applicable, however, to every type of resilient sheet flooring. Unfilled sheet vinyl flooring that has a vinyl backing, for example, is designed to contract after installation and must be installed using special installation procedures and adhesives. Rubber sheet flooring, and sheet vinyl flooring with most types of backing, cannot be heat-welded. In every case, the manufacturer of each sheet flooring material should be contacted and their installation instructions followed.

Sheet flooring should be cut, laid flat, and allowed to acclimate to the temperature of the space in which they will be installed, before installation.

Refer to "Installing Wall Base" later in this chapter for a discussion of separate wall bases. Separate wall base is usually installed after the flooring is in place, where seamless flooring is not required. Where seams are to be welded or sealed, separate wall base should be of the flush-butted type and should be installed before the flooring is laid.

An integral base consists of turning the flooring material up the vertical surfaces. Before an integral base is installed, a continuous fillet (Fig. 4-6) should be placed at the juncture of the floors and the walls. The integral base should be fitted neatly, set in a full bed of adhesive, and rolled. The cove at the floor should be constant. The base material should be in complete contact with the fillet. Absolutely no cracks should exist where the flooring sheet is bent to form the cove radius. A continuous cap (see Fig. 4-6) should be applied at the top of the base.

Sheet flooring should be adhered to the entire floor area in conformance with the manufacturer's printed instructions. Adjacent sheets should be pattern-matched. Where the joints will be welded or sealed, the flooring should be installed in contact with the wall base. The joint between the wall base and the flooring should be heat-welded or sealed using the proper adhesive, to provide a monolithic pinhole-free surface.

The installed sheet flooring should be rolled to ensure a smooth even bond to the substrates, and eliminate air bubbles, ripples, and uneven areas.

Penetrations through the edges of sheet flooring where monolithic flooring is required should be sealed using a sealant furnished by the flooring manufacturer to exclude water.

As few seams as possible should be used. Seams should be located symmetrically in each room or space where they were shown on the layout drawings mentioned earlier in this chapter. Seams should be sharp, straight, and tight. Where monolithic flooring is required, seams should be fully heat-welded or adhesive-sealed. Welded or sealed seams should be pinhole-free. Traces of adhesive and manufacturer's ink markings should be removed from the surfaces of the flooring.

Installing Miscellaneous Resilient Units and Accessories

Installing Resilient Bases, Stair Risers, and Stringers. Refer to "Installing Resilient Sheet Flooring" for a discussion of integral wall bases.

Standard set-on wall bases (see Fig. 4-6) are usually installed after the flooring has been laid. Carpet bases (see Fig. 4-6) are installed before the carpet is placed. Butt-type bases (see Fig. 4-6) are installed before the resilient flooring is laid. Butt-type bases may be used in any installation, but are usually reserved for use with sheet flooring where the joints are to be heat-welded or adhesive-sealed.

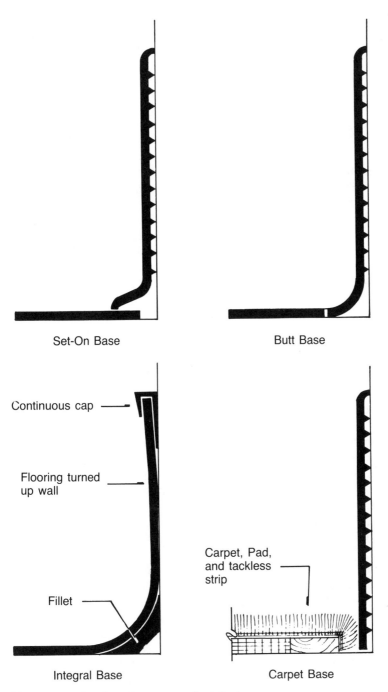

Set-On Base

Butt Base

Continuous cap

Flooring turned
up wall

Fillet

Integral Base

Carpet, Pad,
and tackless
strip

Carpet Base

Figure 4-6 Resilient base types installed.

Base, riser, and stringer material should be unrolled and allowed to acclimate to the temperature and humidity in the space where they will be installed for at least 24 hours before the adhesive is applied.

Stringer material should be cut accurately to fit the stair profile at the wall. Vertical dimensions at the top and bottom of the flight are often the same as the abutting base. A single length should be used for each stringer. The back of the cut stringer material should be entirely covered with adhesive. The material should then be set carefully in place and rolled to ensure a firm bond.

Risers should be cut to fit. After the stair tread has been placed, the back of the riser should be covered with adhesive and the riser placed and rolled to ensure a firm bond.

Wall base corners should be installed before the remainder of the base is installed. Premolded corners should be covered on the back with adhesive and pressed into place. Hand-formed outside corners are produced by making a V-shaped cut at least one-half the thickness of the base or by carving out a 1/4-inch-wide area in the back of the base where the corner will fall, and cutting away a 1-inch-wide area at the heal of the base. Insider corners are made by cutting an inverted V-shape into the front of the cove. The cuts should be made in the center of base pieces which are at least 24 inches long. After the cuts have been made, heat is applied to the cut area with a torch or heat gun until the surface changes to a glossy finish. Then the piece is formed by turning up the cove and bending the sides back to back. The formed piece is then held in position until the material cools, which sets in the new shape. Discolorations created by the forming process are eliminated by reheating after the corner piece has been installed.

Wall base pieces should be set straight, scribed neatly, and fitted tightly to floors, walls, pilasters, columns, casework, cabinetwork, lockers, and other building features. Accurate level and alignment should be maintained. The bottom edge of the base should be in continuous contact with the flooring. The maximum practicable lengths of base should be used. Pieces should never be less than 24 inches long. Joints less than 12 inches from a corner should be avoided, except on pilasters, columns, and similar conditions.

The joints in wall base should be welded when used in conjunction with heat-welded sheet flooring, and adhesive sealed when used with a flooring that has adhesive sealed joints.

Preformed end stops should be used at the exposed ends of wall bases.

Base, riser, or stringer pieces should not be stretched during installation. Where such materials are to be applied to masonry, a coat of mortar should be applied to the masonry to form a smooth surface. The mortar should be allowed to dry thoroughly before the resilient material is installed.

Installing Resilient Stair Treads and Nosings. Treads and nosings should be fit carefully and neatly at risers, ends, and penetrations. Each tread should be covered completely in one piece. Treads should be securely bonded in place using the adhesive recommended by the tread manufacturer.

Installing Resilient Edging (Feature or Accent) Strips. Along the edge where resilient flooring terminates at adjoining exposed concrete floors or other finish flooring, and marble, wood, or metal thresholds, divider strips, or edgings do not occur, resilient edging strips are often applied. Feature strips are also used where resilient floors of different color or pattern occur on opposite sides of openings in partitions; where maintaining the joint pattern in resilient tile through an opening is impractical; and where resilient tile occurs on one side of an opening and sheet flooring occurs on the other side of the opening. Such resilient strips are not, however, normally used where carpet occurs adjacent to the new resilient materials.

The top of the edging strip should finish flush with the resilient flooring. Accurate alignment should be maintained. Joints should be butted tightly.

Excess adhesive should not be left on edging strips or adjoining finish floors or other adjacent surfaces.

At doorways, the edging strips should be mounted so that they lie under the doors when they are closed. The edging strips should be in one piece from jamb to jamb and neatly cut to fit at jambs.

Installing Resilient Reducer Strips. Reducer strips are used where resilient flooring abuts ceramic or quarry tile flooring or any other flooring that is not the same thickness as the resilient flooring at doors and other locations, and marble, metal, or other thresholds do not occur.

At doorways, reducer strips should be installed under the doors when they are closed. The reducers should be in one piece from jamb to jamb and neatly cut to fit at the jambs. They should finish flush with the material on each side of the reducer.

Installing Miscellaneous Accessories. The manufacturer's instruction should be carefully followed when installing miscellaneous resilient accessories.

Cleaning, Protection, and Finishing of Resilient Materials

Excess adhesive should be removed from resilient materials and adjacent surfaces immediately, while the resilient material is being installed. Care should be taken in the cleaning so that surfaces are not damaged.

When the resilient material installation has been completed, adjacent resilient materials and walls, millwork, casework, and other adjacent materials should be thoroughly cleaned of excess adhesive and soiling caused by the installation. Sandpaper, steel wool, or cleaners that will damage surface finishes should not be used.

Installed resilient materials should be scrubbed thoroughly using a non-caustic solution of neutral cleaner, then rinsed thoroughly and allowed to dry.

Unless such finishing is specifically not recommended by the material manufacturer, one thin coat of sealer should be applied to vinyl tile, vinyl composition tile, rubber sheet flooring, rubber tile, and rubber stair treads, using a mop or lambswool applicator and allowed to dry thoroughly. If recommended by the sealer manufacturer, more than one coat of sealer should be used. Then, two coats—more if recommended by the manufacturer—of polish should be applied. The sealer may be omitted if so recommended by the manufacturer of the flooring material and the manufacturer of the polish.

The floors should be thoroughly buffed between coats of polish and after the final coat of polish using a soft-bristle or lambswool buffer.

Vinyl sheet flooring should be sealed and waxed only if the manufacturer recommends doing so. Some sheet vinyl never needs waxing. Some have an abrasive surface texture or an embedded abrasive material, which may have their effect negated if a finish is applied.

After they have been cleaned—and waxed if appropriate—resilient materials subject to damage should be protected during the remainder of the construction period using undyed, untreated, fire-resistant building paper. Planking or plywood walkways should be provided where potential damage due to subsequent construction operations warrants such protection. Stained and damaged resilient materials should be removed and new materials installed.

Resilient Flooring Failures and What to Do about Them

Why Resilient Flooring Fails

Some resilient material failures can be traced to one or several of the following sources: structure failure; structure movement; solid substrate problems; other building element problems. Those sources are discussed in Chapter 2. Many of them, while perhaps not the most probable cause of resilient flooring failure, are more serious and costly to fix than the types of problems discussed in this chapter. Consequently, the possibility that they are responsible for a resilient flooring failure should be investigated.

The causes discussed in Chapter 2 should be ruled out or, if found to be at fault, repaired before resilient flooring repairs are attempted. It will do no good to repair existing, or install new, resilient flooring when a failed, uncorrected, or unaccounted-for problem of the types discussed in Chapter 2 exists. The new installation will also fail. After the problems discussed in Chapter 2 have been investigated and found to be not present, or repaired if found, the next step is to discover any additional causes for the resilient flooring failure and correct them. Possible additional causes include bad materials, improper design, bad workmanship, failure to protect the installation, poor maintenance procedures, and natural aging. The following paragraphs discuss those additional causes. Included with each cause is a numbered list of errors and situations that can cause resilient flooring failure. Refer to "Evidence of Failure" later in this chapter for a listing of the types of failure to which the numbered failure causes apply.

Bad Materials.   Improperly manufactured resilient materials, adhesives, or related items arriving at a construction site is certainly not unheard of, and that possibility should be considered when resilient flooring fails. Since most resilient material imperfections can be easily seen, bad materials should be eliminated before they are installed. At any rate, the number of incidents of bad materials is small compared with the cases of bad design and workmanship. Types of manufacturing defects that might occur include:

1. Resilient flooring materials or accessories that are not true to shape. Examples include square tiles that are actually trapezoids, and stair treads or accessories which are not the same width throughout.
2. Resilient flooring materials with inconsistent colors or patterns.
3. Resilient materials that are stained or otherwise discolored at the factory, or are manufactured using materials that are inconsistent in color.
4. Resilient materials that are damaged in the manufacturing process.
5. Resilient materials that are inconsistent in composition or density.

Improper Design.   Improper design includes selecting the wrong resilient material for the location and conditions, as well as requiring an improper installation. The following design problems can lead to resilient flooring failure:

1. Selecting resilient materials of inappropriate composition for the location and use intended. Using residential non-wax flooring in a commercial lobby, and using rubber flooring in a commercial kitchen, are examples.
2. Selecting adhesives that are of the wrong type for the installation. The wrong adhesives may stay soft, exude from the joints, delaminate from the substrate, not hold the resilient material down, or even dissolve completely.

3. Failing to require proper installation of resilient materials, including requiring that too little adhesive be used, or that the adhesive be improperly applied to the resilient material, or that the resilient material be improperly installed after the adhesive has been applied.
4. Failing to require a concrete topping over a concrete plank floor. Mastic underlayment is not sufficient.

Bad Workmanship. Correct preparation and installation are essential if resilient flooring failures are to be prevented. The following workmanship problems can lead to resilient flooring failure:

1. Failing to follow the design and the recommendations of the manufacturer and recognized authorities.
2. Installing resilient materials before the building has been completely closed and wet work, such as concrete, masonry, and plaster have dried out sufficiently, and where concrete is not dry and at the proper level of alkalinity.
3. Installing resilient materials when the room temperature has been less than 65 degrees Fahrenheit for 48 hours before the installation. Letting the room temperature fall below 65 degrees Fahrenheit within 48 hours after the resilient material application.
4. Failing to properly prepare the area where a resilient material will be installed, removing contaminants and items that will interfere with proper installation, damage the resilient materials, or telegraph through the resilient material.
5. Failing to properly clean the back of resilient material and to prime the substrate before installation.
6. Leaving out underlayment where it is needed. Using the wrong type underlayment. Failing to use latex underlayment when necessary to provide base surfaces which will permit flush connections at edging strips and reducers.
7. Installing damaged resilient materials, regardless of whether the damage was inherent in the manufactured materials or occurred during shipment, storage, or installation.
8. Failing to allow resilient materials to acclimate to the temperature and humidity in the space for at least 48 hours before installing them.
9. Failing to lay cut sheets, base, risers, treads, and stringers flat and allow them to flatten out before installation.
10. Permitting resilient materials to become wet or damp before or during installation.
11. Applying resilient materials to damp concrete or other substrate, or in a damp room.

12. Installing resilient materials that are stained.
13. Installing resilient materials that have damaged edges or faces or broken, cracked, chipped, or deformed materials.
14. Failing to properly install resilient materials, including not using enough adhesive, improperly applying adhesive, and improperly installing the resilient material after the adhesive has been applied. This also includes improperly making adhesive-sealed or heat-welded joints and improperly making field-made wall base corners. Improperly heating or failing to reheat resilient materials can cause discolorations. When a piece of resilient wall base is heated so that it can be formed to fit at a corner, for example, it changes color and may lose its sheen. The color and sheen will return to normal when the material is properly reheated.
15. Failing to match tiles for color and pattern by using tile from cartons in same sequence as manufactured and packaged.
16. Failing to install a complete system, such as leaving out a tile.
17. Using adhesives that are of the wrong type for the installation.
18. Failing to tightly cement the resilient material to the substrate and to press or roll it to eliminate air bubbles and raising or puckering at the joints.
19. Failing to apply a continuous fillet at the juncture of floors and walls for an integral base. Failing to fit the base neatly, set it in a full bed of adhesive, and roll it. Failure to apply a continuous cap at the top of an integral base.
20. Failing to apply a coat of mortar to masonry that will be faced with a resilient wall base or stair stringer.
21. Stretching resilient materials during installation.
22. Failing to make heat-welded and adhesive-sealed seams pinhole-free.
23. Using the wrong finishing material. Applying urethane to a material that should be waxed and waxing a no-wax floor are examples.
24. Improperly repairing an existing floor. Errors range from patching with new materials that do not match the original materials to failure to properly prepare or clean the existing surface.
25. Failing to remove adhesive and manufacturer's ink markings and other soiling from the surfaces of resilient materials.

Failing to Protect the Installation.   Resilient materials must be protected before, during, and after installation. Errors include the following:

1. Failing to protect resilient materials from staining by other construction materials.
2. Permitting abuse before or during installation or after the resilient material has been installed.

3. Permitting the temperature in a space where resilient materials have been installed to fall below 55 degrees Fahrenheit.
4. Covering vinyl or vinyl composition flooring with rubber floor mats, which can permanently stain the flooring.

Poor Maintenance Procedures.   Not maintaining resilient materials properly can result in failures. Poor maintenance procedures include:

1. Failing to clean resilient materials regularly. Unremoved grease, dirt, and other contaminants can result in permanent stains.
2. Improper cleaning of resilient materials. Resilient materials may be damaged severely by sandpaper, steel wool, and abrasive or caustic cleaners.
3. Waxing no-wax floors.

Natural Aging.   Resilient materials may fail due to the following cause:

1. Resilient materials age and lose their properties over time. No-wax vinyl flooring may lose its sheen, for example. Some aging damage may be restored, at least temporarily. Other such damage requires replacing the flooring.

Evidence of Failure

In the following paragraphs, resilient material failures are divided into failure types, such as "Stained or Discolored Resilient Materials." Following each failure type are one or more failure sources, such as "Improper Design." After each failure source, one or more numbers is listed. The numbers represent possible errors associated with that failure source which might cause that failure type to occur.

A description and discussion of the numbered failure causes for failure types "Steel and Concrete Structure Failure," "Wood Structure Failure," and "Structure Movement" appear under those headings in Chapter 2 under the main heading "Structural Framing Systems," subheading "Structure Failure," which are listed in the Contents.

A description and discussion of the numbered failure causes for failure type "Solid Substrate Problems" appear under that heading in Chapter 2 under the main heading "Substrates." Both are listed in the Contents.

A description and discussion of the numbered failure causes for failure type: "Other Building Element Problems" appear under that heading in Chapter 2 under the main heading "Other Building Elements." Both are listed in the Contents.

The following example applies to the failure types listed in the preceding three paragraphs. Clarification and explanation of the numbered cause (2) in the example.

- Solid Substrate Problems: 2 (see Chapter 2).

appears in Chapter 2 under the main heading "Substrates" subheading "Solid Substrate Problems," Cause 2, which reads

2. The solid substrate material cracks or breaks up, joints crack, or surfaces spall due to bad materials, incorrect material selection for the location and application, or bad workmanship.

A description and discussion of the types of problems and numbered failure causes that follow failure types "Bad Materials," "Improper Design," "Bad Workmanship," "Failing to Protect the Installation," "Poor Maintenance Procedures," and "Natural Aging" appears under those headings earlier in this chapter, under the main heading "Resilient Flooring Failures and What to Do about Them," subheading "Why Resilient Flooring Fails," both of which are listed in the Contents. For example, clarification and explanation of the type of problem (Improper Design) and numbered cause (2) in the example

- Improper Design: 2.

appears earlier in this chapter under the main heading "Resilient Flooring Failures and What to Do about Them" subheading "Why Resilient Flooring Fails," sub-subheading "Improper Design," Cause 2, which begins:

2. Selecting adhesives that are of the wrong type for the installation.

Resilient material failure types include the following.

Stained or Discolored Resilient Materials.   Blemishes include stains, discolorations, differences in color or sheen within the resilient material, black marks or other soiling on the surface, and adhesives on the surface. Any one or more of the following may be at fault.

- Solid Substrate Problems: 1 (see Chapter 2).
- Other Building Element Problems: 1, 2 (see Chapter 2).
- Bad Materials: 2, 3.
- Improper Design: 1, 2, 3.
- Bad Workmanship: 1, 7, 12, 14, 15, 17, 23, 24, 25.
- Failing to Protect the Installation: 1, 2, 4.
- Poor Maintenance Procedures: 1, 2, 3.
- Natural Aging: 1.

Damaged Resilient Materials. Damage may include scratched, cracked, or split units, broken edges, depressions in unit faces, holes through units, soft spots, pinholes in welded or sealed joints, or underlying materials telegraphing through the resilient materials. Any one or several of the following may be at fault, depending on the type and extent of the damage.

- Steel and Concrete Structure Failure: 1 (see Chapter 2).
- Wood Structure Failure: 1, 2, 3, 5, 8 (see Chapter 2).
- Structure Movement: 2, 3, 4, 5, 7, 8 (see Chapter 2).
- Solid Substrate Problems: 2, 3, 4 (see Chapter 2).
- Bad Materials: 4, 5.
- Improper Design: 1, 3, 4.
- Bad Workmanship: 1, 4, 6, 7, 13, 14, 19, 20, 22, 24.
- Failing to Protect the Installation: 2.
- Poor Maintenance Procedures: 2.

Missing, Loose, Raised, Curled, or Out-of-line Resilient Materials, and Materials Having Blisters or Air Bubbles. The damage may be due to one or more of the following causes.

- Steel and Concrete Structure Failure: 1 (see Chapter 2).
- Wood Structure Failure: 1, 2, 3, 5, 8 (see Chapter 2).
- Structure Movement: 2, 3, 4, 5, 7, 8 (see Chapter 2).
- Solid Substrate Problems: 1, 2, 3, 4 (see Chapter 2).
- Other Building Element Problems: 1, 2 (see Chapter 2).
- Bad Materials: 5.
- Improper Design: 2, 3, 4.
- Bad Workmanship: 1, 2, 3, 4, 5, 6, 8, 9, 10, 11, 13, 14, 16, 17, 18, 19, 21, 24.
- Failing to Protect the Installation: 2, 3.

Poorly Fitting Resilient Materials. Resilient materials may have open joints or fit poorly against adjoining surfaces for one of the following causes.

- Steel and Concrete Structure Failure: 1 (see Chapter 2).
- Wood Structure Failure: 1, 2, 3, 5, 8 (see Chapter 2).
- Structure Movement: 2, 3, 4, 5, 7, 8 (see Chapter 2).
- Solid Substrate Problems: 3, 4 (see Chapter 2).
- Bad Materials: 1.
- Bad Workmanship: 1, 6, 7, 8, 9, 13, 14, 19, 20, 21.

Cleaning, Repairing, and Refinishing Resilient Flooring

General Requirements

The extent of cleaning, repairing, and refinishing of resilient materials depends on the type and extent of the damage. This section discusses the cleaning and refinishing of existing materials to remain exposed, and repairs needed when the damage is minor. "Installing New Resilient Flooring over Existing Materials" later in this chapter discusses major extensions of existing resilient materials, preparation of an existing resilient floor that will be covered by new resilient flooring or carpet, and the installation of new flooring over an existing resilient floor.

The following paragraphs contain some suggestions for cleaning, repairing, and refinishing resilient materials. Because the suggestions are meant to apply to many situations, they might not apply to a specific case. In addition, there are many possible cases that are not specifically covered here. When a condition arises in the field that is not addressed here, advice should be sought from the additional data sources mentioned in this book. Often, consultation with the manufacturer of the materials being cleaned, repaired, or refinished will help. Sometimes, it is necessary to obtain professional help (see Chapter 1). Under no circumstances should the specific recommendations in this chapter be followed without careful investigation and the application of professional expertise and judgment.

Before an attempt is made to clean, repair, or refinish existing resilient materials, the existing materials manufacturer's brochures for products, installation details, and recommendations for cleaning, repairing, and refinishing should be available and referenced. It is necessary to be sure that the manufacturer's recommended precautions against the use of materials and methods that may be detrimental to the materials are followed.

It is virtually impossible to remove properly installed resilient materials without damaging them to the extent that reuse is impossible. Therefore, when this chapter uses the word repair with regard to resilient material, it almost always means removing the material and installing new material.

A check should be made to see whether sufficient extra material was left from the original installation to make the repairs. Often as much as three percent of the amount of resilient material actually installed will be left with the owner for making future repairs.

After resilient materials have been removed, concealed damage to the underlying construction may become apparent. Such concealed damage should be repaired before new resilient materials are installed.

Existing resilient materials that are to be removed should be removed carefully, and adjacent surfaces should be protected so that the process does no damage to the surrounding area or people in the vicinity. Since the existing lining felt, underlayment, adhesives, or resilient materials may contain asbestos or other harmful substances, they should never be sanded or handled in any other way that will cause dust or loose particles to be released into the air. Some sources recommend that existing resilient materials be removed only when absolutely necessary, and then only by people trained in removing hazardous materials. They say that encapsulation, such as by covering with an underlayment and new flooring, is a better solution.

Areas where repairs will be made should be inspected carefully to verify that existing materials that should be removed have been removed, and that the substrates and structure are as expected and are not damaged. Sometimes, substrate or structure materials, systems, or conditions are encountered which differ considerably from those expected. Sometimes unexpected damage is discovered. Both damage that was previously known and damage found later should be repaired before the resilient flooring is repaired. Cleaning and refinishing should be done, substrates should be prepared, and the repairs made strictly in accordance with the material manufacturer's recommendations and the recommendations of the Resilient Floor Covering Institute. Repairs should be made only by experienced workers.

Materials to be used for making repairs should not be permitted to freeze. Spaces should be kept above 55 degrees Fahrenheit before and after the repairs are made.

Adequate ventilation should be provided while repairs are being made, to ensure the safety of the workers. The resilient material manufacturer's recommended safety precautions should be followed.

The moisture content of the substrates should be within the range recommended by the resilient materials manufacturer, especially if excess moisture contributed to the failure being repaired.

Structure Framing, Substrates, and Other Building Elements.   While improperly designed or installed, or damaged steel, concrete, or wood framing, solid substrates, or other building elements, which are discussed in Chapter 2, may be responsible for resilient flooring failure, repairing them is beyond the scope of this book. They should be investigated as possible sources of the failure, and repaired as necessary. The discussions in this chapter assume that when those items have been the cause of a failure, they have been repaired and present satisfactory support systems and conditions for the materials being repaired.

Materials

Resilient Materials.   Repairs should be made using resilient materials that exactly match the existing materials. Care must be exercised when selecting materials from a different manufacturer than the one that made the materials originally used. Products of the same type, pattern, and color, may vary significantly in appearance.

A check should be made to determine whether sufficient materials were left from the original installation to complete the repairs.

Adhesives, Welding Rods, and Other Miscellaneous Materials.   All miscellaneous materials needed for the repairs should be types suitable for the work to be done. Such materials must, of course, be compatible with the new and existing materials with which they will come into contact.

Cleaning Materials.   In general, the same cleaning products mentioned earlier in this chapter under ''Cleaning and Polishing Materials,'' which is listed in the Contents, for use on new floors, are used to clean existing floors. Additional materials needed include the following:

- Cleaners for existing floors: For everyday stains, a good approach is to start with the least caustic material available and work up toward more potentially harmful substances. Common substances that will frequently remove stains include, in the order of the increasing possibility of harm to the resilient material, water, soap and water, white vinegar and water, hydrogen peroxide, rubbing alcohol, household ammonia, lighter fluid, and nail polish remover. As we will discuss further on in this chapter, however, materials containing grease or solvents should never be used on rubber resilient materials. Where common materials are not successful, a fast-acting cleaner formulated to clean tough dirt and grime may be used. Cleaners should be pH safe for resilient floors, and of a type that will leave no dulling soap scum build-up.
- Stripping solution: Complete removal of existing finishes requires the use of a stripping solution. It should be of a type and brand recommended by the resilient material, sealer, and polish manufacturers, to ensure that it does not damage the flooring. Stripping solutions are formulated for removing the applied finish from discolored resilient flooring and resilient flooring with heavy polish build-up. The stripping solution may differ depending on the type of polish to be removed.

Cleaning and Refinishing Resilient Materials

The manufacturer's instructions should be obtained and followed for the cleaning and refinishing of resilient materials. The following suggestions are general in nature and should not take precedence over the manufacturer's instructions.

Dust and loose dirt, especially gritty dirt, should be removed daily using a dust mop or by sweeping. Substances that will stain the resilient materials should be removed immediately, using a clean white cloth and a commercial detergent floor cleaner. Already present stains may be removed using one of the materials mentioned earlier for cleaning existing floors.

Periodically, resilient materials should be damp-mopped with a mild detergent solution, or gone over with a machine using a spray-buff solution to remove black scuff marks and foreign materials such as chewing gum. After the damp-mopping, most floors should be machine buffed. Some slip-resistant floors, however, should never be buffed.

Occasionally, floors that were originally covered with floor polish should be given a new coat and then buffed after the polish has dried. Most of the time, new polish is only necessary in areas where the traffic is heavy.

Floors that have been allowed to accumulate a heavy polish build-up should be stripped using a commercial stripping solution, and refinished in the same manner as was discussed earlier for new floors.

Oils, grease, sweeping compounds that contain grease, and treated mops should not be used to clean rubber flooring. Solvents such as gasoline, kerosene, and naphtha should not be used on rubber. Such materials should be used on other resilient materials only with the consent of the resilient material's manufacturer.

To prevent abrasive burns caused by high-speed buffing machines, floors should be lubricated during buffing using water or a buffing solution. Buffing pads should be changed frequently.

There is a method in use today called "no-wax" maintenance. In this method, the surface of the resilient flooring is buffed without the use of polishes or lubricants. The process melts the surface of the flooring and causes a sheen to appear as small scratches disappear. This process yields a satisfactory floor finish at least temporarily, but some manufacturers recommend that the process not be used. They say that small particles of dirt are bound into the resilient surface, eventually darkening the material, and that the process may cause the flooring to wear out faster than if more conventional cleaning and polishing methods are used.

Repairing Resilient Materials

Removing Existing Resilient Materials.   The first step in preparing to make repairs is to remove the damaged resilient materials. Where portions of the existing material are removed, it is normally necessary to also remove all traces of old adhesive, oil, wax, dust, paint, and other materials that would damage new resilient materials; prevent proper application of new resilient materials, carpet, or other new materials; or reduce an adhesive bond. It is sometimes permissible to leave some residue of existing adhesive, however. Another exception to the full-removal rule is sometimes made for some types of sheet flooring that have a backing or felt layer. In those cases, the surface only is removed. The backing is left in place to serve as a base for some types of new sheet flooring. The suitability of leaving existing materials in place when resilient materials are removed must be verified with the resilient materials manufacturer.

When removing resilient materials, the recommendations of the manufacturer and the Resilient Floor Covering Institute should be carefully followed. Sometimes adhesives can be removed using solvents such as lacquer thinner, methyl ethyl keytone (MEK), and others. Such solvents must be used with care, of course. Many of them are flammable, caustic, or both. Unless it is positively proven by laboratory analysis that asbestos and other harmful substances are *not* present in the resilient material, underlayments, lining felt, and adhesive, no dust-producing methods should ever be used to remove any of those materials. Where scraping is necessary, only wet scraping and wet sweeping should be used. Materials should be removed in one piece where possible and disposed of without being broken up. Existing materials should not be sanded. Other precautions are also most likely necessary.

Protecting Resilient Materials That Remain in Place. When portions of an existing resilient material application are damaged or must be removed for other reasons, the portions of the material that remain in place should be protected to minimize damage while repairs or other work in the space are in progress. Usually, a paper covering is sufficient to protect remaining materials. But sometimes, when heavier than normal foot traffic or wheeled travel is expected, for example, it is necessary to place a layer of plywood over the remaining resilient flooring.

To minimize the need for additional repairs, only the damaged portion of resilient materials should be removed. Often, the process of removing such materials will damage them to the extent that they are not suitable for reuse.

Preparing Existing Substrates to Receive Patches. Substrates should be made sound, smooth, and clean. The printed recommendations of the materials manufacturer should be followed.

In wood substrates, loose boards should be renailed, defective and badly worn boards should be removed, and the resulting openings filled with new materials. Cracks and holes should be filled with plastic wood or snugly fitting wood pieces. Surface irregularities, such as cupping, should be removed by sanding. The requirement to refrain from sanding resilient materials should be observed, of course. When the wood substrate is in a condition that is too poor to be repaired using normal means, a board underlayment may be necessary. It is, of course, not possible to install a board underlayment over only a portion of a wood substrate. Situations where an underlayment are required are discussed in "Installing New Resilient Flooring over Existing Materials" later in this chapter.

It is necessary to remove from concrete substrates all traces of paint, grease, oil, wax, mortar, old adhesives, floor coverings, floor hardener, and other foreign substances that would reduce bond or harm new materials. Uneven or rough surfaces should be repaired using a latex-modified cement underlayment recommended by the resilient material manufacturer for the purpose. The concrete should be in a condition equal to that required for new concrete. Concrete substrates should be tested for moisture content, as discussed earlier in this chapter for new concrete. The concrete should be primed if so recommended by the resilient material manufacturer.

Other substrates should be prepared in accordance with the resilient materials manufacturer's recommendations.

Adjoining surfaces to receive resilient materials should be in the same plane. Adjacent surfaces to receive different finish floor materials should be in proper planes to permit acceptable transitions. Materials and methods should be used that will produce smooth surfaces in the proper plane.

Surfaces to receive resilient materials should be clean and dry before resilient material application is started. They should be swept or vacuum cleaned immediately before the resilient materials are installed.

Repairing Existing Resilient Materials.   The requirements for installing new materials on new substrates, as discussed earlier in this chapter, generally apply to installing patches. For example, only proper adhesive of the type and brand recommended by the resilient materials manufacturer should be used, and they should be used without adulteration or reduction, in accordance with the manufacturer's instructions. Only as much adhesive should be applied as can be covered with resilient material within the recommended working time of the adhesive.

New materials used in making repairs should be cut, fit, and scribed to fit neatly where existing materials have been removed. Joints should be butted tightly. Installed materials should match those already in place.

The method used to repair resilient materials depends on the type and extent of damage. Some minor scratches, for example, can be buffed away. Scratches and dents in materials definitely known to not contain asbestos can sometimes be repaired using scraps of the resilient materials. Some material is scraped from the surface of the non-asbestos-containing scrap materials and ground to a powder. The powder is mixed with a colorless lacquer or quick-drying varnish to form a paste. The paste is applied to the damaged area, allowed to dry, and buffed with fine steel wool and boiled linseed oil. The oil is wiped off, and the repaired resilient material waxed.

This method will not work, of course, on materials that will be damaged by the materials used. Larger nicks, gouges, and burns can probably not be removed from most resilient materials.

No-wax vinyl flooring will lose its sheen eventually. When that happens, those products that were made no-wax by an applied coating can be restored using a special top dressing available from the manufacturer. Where the no-wax feature in inherent in the wearlayer, buffing may restore the no-wax look.

Cracks; curled edges and other deformations; missing materials; and soft, permanently tacky, crumbling, or otherwise badly damaged surfaces are all major damage. Sometimes materials with curled edges can be straightened. A hot iron or heat lamp is used to soften the adhesive holding the curled material in place. When the adhesive is soft, the resilient material is lifted, a dab of adhesive is shoved beneath the curled edge and the resilient material flattened out into the adhesive. A weight is applied until the adhesive has completely dried. Usually, however, such major damage cannot be repaired. The damaged materials must be removed and similar materials installed.

Where existing resilient flooring is to be left in place and not covered with another flooring, new resilient materials should be installed where existing damaged materials were removed and where materials are missing for other reasons.

The new materials installed where existing material is missing should exactly match the adjacent materials in every way, including type, material, size, thickness, color, and pattern. Patches should be inconspicuous.

When new tile material is available in type and color but not in size to match the existing material, minor repairs can be made by cutting larger-size tile to match the existing tile.

When new material is not available to match the existing material, it may be necessary to completely remove the existing resilient material in the entire space and install new acceptable materials.

When the damage is major in scope (a subjective judgment), it may be better to completely remove the existing resilient material in the entire space and provide new acceptable material. An alternative would be to patch the existing material level using compatible resilient materials, cover it with a board underlayment, and install a new layer of flooring. Refer to "Installing New Resilient Flooring over Existing Materials" later in this chapter for a discussion of installing new resilient flooring over existing resilient flooring.

Existing resilient wall base should be completely removed when adjacent resilient flooring to be removed lies beneath the base. Damaged sections of resilient wall base should not be cut out between existing joints. Instead, the entire section of which a portion is damaged should be removed.

Damaged resilient stair treads, risers, and stringers, and accessories such as reducers, thresholds, and feature strips should be removed entirely.

Resilient materials should be installed smoothly, without air pockets, in accurate alignment, with joints brought in tight contact, and without raising or puckering at joints, telegraphing of adhesive spreader marks through the material, or other surface imperfections. Resilient materials should be tightly cemented to the substrate.

New materials should be matched as closely as possible with the material left in place.

Latex underlayment should be used as necessary to provide surfaces which will permit flush connections as adjoining materials.

Broken, cracked, chipped, or deformed resilient materials should not be used in making repairs.

Tile units should be laid in the same fashion, with the grain reversed in alternate units or with the grain in all tile running in same direction, as was done in the original installation. Unless inappropriate, joints should align with those in the original installation.

Rolled materials should be unrolled and allowed to flatten out and acclimate to the space before they are installed.

Sheet flooring should be pressed in place with a roller to ensure a smooth, even bond to the substrate, eliminate air bubbles, ripples, and uneven areas. A constant radius should be created at the wall base cove to match the existing condition. Sheets should be solidly in contact with cove fillet strips. Absolutely no cracks should exist where sheets are bent to form a cove radius. Caps should be provided at the top of an integral base. Seams should be installed in a sharp, straight, and tight-fitting manner and either left unsealed, sealed using an adhesive, or heat-welded, as was done in the original application. Sealed seams should create monolithic surfaces completely free from pinholes. Adjacent sheets should be pattern-matched.

Where base or stringer material is to be applied to masonry, and a coat of mortar has not been applied to form a smooth surface to receive the resilient material, such a mortar coat should be applied.

The same type of wall base as was used in the original installation, including the same type of corner sections, should be installed where the existing base has been removed. New base sections should be set straight, in alignment with existing base sections that were left in place, scribed neatly, and fit tightly to floors, walls, pilasters, columns, casework, and other building features. Accurate level and alignment should be maintained. The bottom edge of the new base sections should be in continuous contact with the flooring.

Resilient materials used in repairs should not be stretched during installation.

Traces of excess adhesive and manufacturer's ink markings should be removed from surfaces of resilient materials and adjacent materials.

Cleaning, Protecting, and Finishing Repaired Resilient Materials. Repaired surfaces should be cleaned, finished, and protected in the same way as would be a new application. Refer to "Cleaning, Protection, and Finishing" earlier in this chapter. It is usually necessary to completely remove existing applied finishes. Removing only a portion of them will often leave splotchy surfaces. Applying a new finish to only the repaired portion will probably be apparent. Removing an applied finish is discussed earlier in this chapter under "Cleaning and Refinishing Resilient Materials."

Installing New Resilient Flooring over Existing Materials

Requirements Common to All Resilient Flooring

The following requirements contain some suggestions for installing new resilient flooring in existing spaces. Because the suggestions are meant to apply to many situations, they may not apply to a specific case. In addition, there are many possible cases that are not specifically covered here. When a condition arises in the field that is not addressed here, advice should be sought from the additional data sources mentioned in this book. Often, consultation with the manufacturer of the materials being installed will help. Sometimes it may be necessary to obtain professional help (see Chapter 1). Under no circumstances should these recommendations be followed without careful investigation and the application of professional expertise and judgment.

The discussion earlier in this chapter about new resilient flooring materials and installations applies to new resilient flooring installed in existing spaces. There are, however, a few additional considerations to address when new flooring is installed in an existing space. The following paragraphs address these concerns. Some, but not all, of the requirements discussed earlier are repeated here for the sake of clarity. It is suggested that the reader refer to earlier paragraphs for additional data.

The requirements listed earlier in this chapter under "Resilient Flooring Materials," "Resilient Flooring Installation," and "Cleaning, Repairing, and Refinishing Resilient Flooring" apply. Substrates should be prepared and new materials applied in accordance with the resilient flooring material manufacturer's recommendations. Only experienced workers should be used to install resilient materials.

Materials manufacturer's brochures, including installation, cleaning, and maintenance instructions, should be collected for each type of resilient material, adhesive, underlayment, cleaning, and finishing material and kept available at the site.

A working layout should be prepared for each area to receive resilient flooring. It should show the location of each seam, location and type of bases, edgings, and accessories, changes in color, and other pertinent installation details.

Adequate ventilation should be provided while work associated with resilient materials preparation and application are in progress. The resilient materials manufacturer's recommended safety precautions should be followed.

Existing resilient materials that are to be removed should be removed carefully, and adjacent surfaces should be protected so that the process does no damage to the surrounding area or people in the vicinity. Since the existing lining felt, underlayment, adhesives, or resilient materials may contain asbestos or other harmful substances they should never be sanded or handled in any other way that will cause dust or loose particles to be released into the air. Some sources recommend that existing resilient materials be removed only when absolutely necessary, and then only by people trained in removing hazardous materials. They say that encapsulation, such as by covering with an underlayment and new flooring, is a better solution.

Areas where repairs will be made should be inspected carefully to verify that existing materials that should be removed have been removed, and that the substrates and structure are as expected and are not damaged. Sometimes substrate or structure materials, systems, or conditions are encountered which differ considerably from those expected. Sometimes unexpected damage is discovered. Both damage that was previously known and damage found later should be repaired before new resilient material is installed.

Structural Framing, Substrates, and Other Building Elements. While improperly designed or installed, or damaged steel, concrete, or wood framing, solid substrates, or other building elements, which are discussed in Chapter 2, may occur, repairing them is beyond the scope of this book. The discussions in this chapter assume that those items have been repaired, if necessary, and present satisfactory support systems and conditions for the new flooring being installed.

Materials. Tile, sheet materials, other units, accessories, underlayments, adhesives, cleaning and finishing materials, and other materials should be as listed in the part of this chapter called "Resilient Flooring Materials." Additional materials required when existing flooring is present include the cleaner for existing flooring and stripping solutions listed earlier in this chapter under "Cleaning, Repairing, and Refinishing Resilient Flooring," subheading "General Requirements," sub-subheading "Materials."

New materials should be selected for their compatibility with existing materials that will remain in place. Materials selected should include resilient flooring; bases; stair treads, risers, and stringer covers; accessories; adhesives; and underlayments. Tests should be conducted in the building to ensure compatibility. Incompatible materials should not be used.

Extra material of same type and colors as those installed should be delivered and stored at the site when the resilient materials work has been completed, for the owner's use in repairing and maintaining the resilient work. The quantity of extra materials should be about 3 percent of the actual amount used in the work. Usually not less than one carton of tile or one roll of sheet flooring is left. Additional wall base; stair treads, riser, and stringer materials; accessories; adhesives; joint materials; and other items should also be left. The quantity should be sufficient to make future repairs.

Materials, including extra materials, should be delivered to the project site in the manufacturer's original packaging. Materials should be packed, stored, and handled carefully to prevent damage.

Removing Existing Flooring.   Complete removal of existing resilient flooring is often not required in areas to receive new resilient flooring or carpet. When the existing resilient flooring is smooth, well-bonded, and compatible with the new adhesives to be used, and its condition is otherwise satisfactory, applying a new flooring directly over an existing resilient material is often possible. Even when it has sustained minor damage, existing resilient flooring in spaces that are to receive new resilient flooring or carpet is often left in place and patched or repaired. Even when some kinds of major damage are present, it may be possible to patch the bad areas and install underlayment over the existing flooring. Sometimes, it will make sense to completely remove severely deteriorated portions of the existing flooring and patch the remainder. In those cases, the new flooring manufacturer may recommend that the entire floor, both patched and new, be covered with an underlayment. In many cases, cost is the deciding factor in determining whether to remove, patch, or provide underlayment.

There are some cases, however, where the decision is dictated by other considerations. For example, unless the manufacturer of the new flooring specifically agrees that it may be left in place, existing resilient material should be completely removed when the substrate is wood; when the new flooring will be ceramic tile or a cementitious product; and when a sufficient amount of the resilient material is damaged, loose, or hollow-sounding as to make patching uneconomical.

When asbestos is present in the existing materials, that alone may be sufficient to dictate that an existing floor be left in place and an overlayment provided, rather than that the flooring be removed.

Unless the manufacturer of the new flooring specifically agrees that the new flooring may be applied directly to it after minor patching, existing resilient flooring should be either completely removed, or patched and covered with an underlayment when the resilient material is badly cracked, blistered, or resists all attempts to re-adhere curled edges and peaks. A good axiom to heed is that, if an existing resilient floor looks unsatisfactory as an underlayment for new flooring, it probably is. Resilient material to which new flooring is to be directly applied must be in good enough condition for the new adhesives to bond; able to receive tackless strips (where the new flooring is carpet); and able to resist the new loads to be applied. Special consideration should be given to the existing material's ability to satisfactorily sustained new rolling loads. Borderline problems should not be covered with new flooring until repairs have been made. Failing to properly prepare an existing resilient material before installing a new material over it may eventually lead to failures.

Removals should be performed so that new finishes are flush with adjacent like finishes, and so that transitions to other finishes are acceptable.

When existing flooring is removed, it is usually necessary to completely remove all existing resilient materials and all traces of old adhesive, oil, wax, dust, paint, and other materials that would harm the new flooring; or prevent proper application of new resilient flooring, carpet or other new materials; or reduce the adhesive bond. Some existing adhesives may not be compatible with the new materials to be applied. Resin-based resilient flooring adhesives will react unfavorably with latex carpet adhesives, for example. Linoleum adhesive is not compatible with some carpet adhesives. The existing adhesive may have been the wrong type for the location. All adhesives should not be used on slabs-on-grade, for example. Even so, in some cases, it may be permissible to leave portions of the original adhesive in place. Sometimes, even the backing on a sheet vinyl flooring material may be left as a base for new sheet vinyl flooring. In both of the latter cases, it is necessary to verify the suitability of the action with the flooring manufacturer.

Even where existing flooring does not need to be removed completely, it may be necessary to remove portions of the resilient materials to perform other work or to make repairs to the flooring. Refer to "Repairing Resilient Materials" earlier in this chapter for information about patching existing resilient flooring that will remain exposed as resilient flooring.

Resilient flooring which is left in place should be protected using plywood or paper, as necessary, to minimize damage.

Existing flooring which is not removed but will be covered by a new material can be handled in one of two ways. It may have minor damage, either existing or caused by construction operations, that can be patched to present a level surface. Openings should be filled, including those left by the removal of existing flooring, walls, or partitions, and those existing at the start of work. The material used in making patches should match the existing flooring in material and thickness. Alternatively, such existing flooring may be completely removed, in lieu of patching. The decision will probably depend mainly on costs. Where repairs to such existing resilient flooring are major, it may be less costly to completely remove it from the entire space. The presence of asbestos, on the other hand, may increase considerably the costs of removal.

Removing Existing Resilient Wall Base. It is usually best to remove existing wall base wherever it occurs in an area to be altered or remodeled. Even where removal is not necessary, existing base that is damaged either when the work starts or during the construction process should be removed and discarded, and new base installed. Only the damaged sections of base need be removed under those circumstances, but the entire damaged section should be removed. Existing base should not be cut between the existing joints.

Where existing base is removed, unless the substrate is also removed, or furred for a new finish, all traces of adhesive, oil, wax, dust, paint, and other materials that would affect an adhesive bond or harm the new materials must also be removed.

Removing Other Resilient Materials. Where new similar materials are to be installed, existing resilient stair treads, risers, and stringers must be removed. Feature strips may be left in place unless they interfere with the new material. Reducers and thresholds will probably have to be removed. Where such materials are removed, unless the substrate is also removed, or furred for a new finish, all traces of adhesive, oil, wax, dust, paint, and other materials that would affect an adhesive bond or harm the new materials must also be removed.

Preparation of Existing Substrates. The following recommendations apply to substrates to receive new resilient flooring that were exposed in the existing building, and those that become exposed when existing flooring is removed. When preparing substrates to receive resilient materials, the printed recommendations of the resilient materials manufacturer should be followed.

Adjoining surfaces to receive resilient materials should be in the same plane. Surfaces to receive different finish floor materials shall be in proper planes to permit acceptable transitions. Whatever methods are needed should be used to produce smooth surfaces in the proper plane.

Surfaces to receive resilient materials must be clean and dry before resilient material application is started.

Surfaces to receive resilient materials should be swept and vacuumed clean immediately before the resilient materials are installed.

Preparation of Wood and Plywood Substrates. Severely damaged wood and plywood flooring, including flooring that is rotted or otherwise not structurally sound, moving and not easily refastened, or severely warped or otherwise misshapen, should probably be removed and an entirely new floor installed. A change in use that requires the floor to carry much increased loads may dictate that a new floor be provided. Installation of a completely new floor is beyond the scope of this book, however.

New resilient flooring may be installed directly over properly prepared plywood or tongue and groove wood strip flooring with boards that are 3 inches or less in width. Proper preparation includes renailing loose boards, removing defective or badly worn boards, and filling resulting openings with new materials. Knot holes, cracks wider than 1/16 inch, and holes larger than 3/16 inch in diameter should be filled with snugly fitting wood pieces, plastic wood, or a crack filler recommended by the resilient flooring manufacturer. Surfaces should be planed, scraped, or sanded smooth to remove surface irregularities, such as cupping. Nails should be set. A 15-pound asphalt lining felt may be recommended by the resilient flooring manufacturer. If so, the felt should be cut to fit at vertical surfaces. Seams should be butted. Cross seams should be staggered. A suitable adhesive should be spread on the wood floor and the felt rolled into the adhesive with a suitable roller weighing about 100 pounds, to ensure that air bubbles will be removed and that the felt will be fully adhered in place.

Wood strip flooring which is not in bad enough condition to require complete removal, but which has boards that are in such condition that sanding cannot eliminate irregularities sufficiently to receive a satisfactory resilient flooring installation, should be covered with an underlayment. When such wood flooring is tongue and groove and has boards 3 inches or less in width, it may be covered with 1/4-inch-thick underlayment board. When such wood flooring is not tongue and groove and has boards that are not more than 3 inches wide, or is tongue and groove but is in poor condition, it should receive a 1/2-inch or thicker plywood underlayment. Wood flooring that has boards 6 inches or more in width, or which is badly cupped, should receive a 3/8- to 5/8-inch-thick plywood underlayment. Lining felt is not usually needed when a plywood underlayment is used.

Where board underlayment is to be used, loose boards should be renailed, defective or badly worn boards should be removed, and resulting openings filled with new materials. Cracks and holes should be filled with plastic wood or snugly fitting wood pieces. Surface irregularities, such as cupping, should be removed by sanding.

Plywood underlayment should be installed in accordance with American Plywood Association (APA) recommendations in the APA Data File "Installation and Preparation of Plywood Underlayment for Thin Resilient (Non-textile) Flooring." The top surface of board underlayments should be level where contiguous.

As discussed earlier, a wood or plywood floor built over concrete, even when sleepers are used, is not an acceptable base for resilient flooring.

Under some circumstances, which should be verified with the resilient flooring and underlayment manufacturers in every case, self-leveling cementitious underlayments may be used over essentially sound existing wood substrates.

Preparation of Concrete Substrates.   There are two ways to prepare an existing concrete substrate to receive new resilient flooring. One way is to clean it to remove all traces of paint, grease, oil, wax, mortar, old adhesives, floor coverings, floor hardener, curing compounds, and other foreign substances, that would reduce bond or harm new materials. The second way is to cover the concrete with an underlayment.

When the existing finish can be easily removed and the concrete is sound, level, and fairly smooth, the first method is probably the better. After harmful materials have been removed, uneven places should be leveled by chipping, filling, or grinding. Minor construction joints, grooves, cracks, holes, and rough areas should also be filled and smoothed out with a latex-modified underlayment, to provide uniform surfaces as recommended by the resilient material manufacturer. The surfaces should be left in a condition equal to that required of new concrete. After the latex underlayment has dried, concrete primer should be properly applied when so recommended by resilient materials manufacturer.

When the concrete substrate is rough, not level, covered with a finish that will result in a rough surface when it has been removed, or has an existing underlayment that has become unsound, a new underlayment or concrete topping may be the better solution. Cases where the existing flooring is a resilient material are discussed later in this chapter. A lightweight concrete topping is, of course, a possibility, especially if the variations in the existing construction are large. Often, though, a self-leveling cementitious underlayment will offer a successful alternative. The manufacturers of the flooring and the underlayment should be consulted for their specific recommendations about the particular case. The new topping or underlayment should be primed when recommended by the resilient flooring manufacturer.

Whatever method is used to prepare the existing concrete, it should be tested for moisture content, as outlined earlier in this chapter for new concrete. When the existing concrete's moisture level exceeds that recommended by the manufacturer of the resilient materials, or the manufacturers of the underlayments or adhesives to be used, the concrete should be caused to dry properly before installation of the resilient flooring is attempted, or floor covering materials should be selected which are more compatible with the conditions.

Preparation of Existing Resilient Flooring.   Existing resilient flooring that has been left in place, and which will be covered with new resilient flooring, should be patched and repaired as necessary to produce a smooth and sound surface in the same plane throughout. Patches should be made using resilient flooring of the same composition and thickness as the existing material.

Existing resilient flooring to remain and be covered by other materials should be stripped. Stripping should remove all old polish build-up and existing sealers. The stripping solution should be used in strict accordance with its manufacturer's recommendations. Care should be taken to prevent damaging the flooring. Flooring damaged by stripping should be removed and new matching flooring installed or completely removed and the substrate cleaned.

General Installation Requirements.   Some requirements are applicable to all types of resilient materials. For example, materials should not be permitted to freeze. Materials and spaces should be maintained at not less than 65 degrees Fahrenheit for a minimum of 48 hours before until 48 hours after installation, and above 55 degrees Fahrenheit after that.

Installation work should not be started in a space until the work of other trades, including painting, has been substantially completed and, during winter months, not until the permanent heating system is operating. The moisture content of concrete, building air temperature, and relative humidity must be within the limits recommended by the flooring manufacturer.

An inspection should be made to verify that surfaces to receive resilient materials are in the proper condition to begin work. Installation should not be started until unsatisfactory conditions have been properly corrected.

Only accepted, proper adhesive, without adulteration or reduction, should be used in accordance with the manufacturer's instructions.

Adhesives should be applied only to as much area as can be covered with resilient material within the recommended working time of the adhesive.

New flooring should be cut, fit, and scribed to fit neatly at walls, breaks, recesses, pipes, door frames, cabinets, fixtures, equipment, thresholds, and edgings. Flooring should be butted tightly to other materials and surfaces.

Installing Tiles, Sheet Flooring, Other Units, Accessories, and Miscellaneous Materials

Specific requirements delineated earlier in this chapter under "Resilient Flooring Installation" apply to resilient tile, sheet flooring, other units, and accessories installed over existing materials, and the adhesives and other materials used to install them.

Cleaning, Protection, and Finishing

The same materials and methods discussed earlier in this chapter under "Resilient Flooring Installation," subheading "Cleaning, Protection, and Finishing" for cleaning, protecting, and finishing new resilient materials in new buildings are used when the resilient materials are installed over existing materials.

Where to Get More Information

The Resilient Floor Covering Institute's technical information is a good source of information. Ask for *SV-1 Resilient Floor Covering Institute Recommended Specification for Resilient Floor Covering—Vinyl Plastic Sheet, Recommended Work Procedures for Resilient Floor Coverings,* and any other data RFCI currently has available.

The American Plywood Association publications APA Source list "Plywood Underlayment for Use under Resilient Finish Flooring" and Data File "Installation and Preparation of Plywood Underlayment for Thin Resilient (Non-textile) Flooring" are very useful when board underlayment is required.

Masterspec's Section 09650 "Resilient Flooring" contains a good discussion of resilient flooring in general, though rubber sheet, linoleum, and cork flooring are not included and only certain types of other flooring are included. Unfortunately, Masterspec section 09650 does not discuss repairs or maintenance.

Unfortunately, the best sources of information about maintaining and repairing resilient materials and installing them over existing materials are the manufacturers of the materials. While some of the data some of them publish is very helpful and accurate, all manufacturer's recommendations do not agree and some manufacturer's data is not as complete as others. Since the recommendations of the particular manufacturer whose product is involved should be followed, it is sometimes hard to get good advice. A manufacturer whose product is not involved cannot be expected to give advice concerning its competitors' product, although they sometimes will do so. The manuals produced by Armstrong are among the most complete in the industry and would be a helpful addition to the library of anyone who deals with resilient flooring.

The referenced U.S. General Services Administration Federal Specifications should be available to anyone selecting or verifying the resilient material used. Those required include L-F-475, L-F-001641, RP-T-650, SS-W-40, SS-T-312, and ZZ-T-001237. Federal Specification ZZ-F-461, which is mentioned in the text and referenced by some rubber sheet manufacturers as applicable to their products, was canceled in 1965 and may be difficult or impossible to find today.

The referenced ASTM Standards E 84, E 648, and E 662, while interesting, all describe test methods for fire resistance characteristics and are probably not essential to someone dealing with the repair or maintenance of resilient materials or the installation of them over existing materials. Designers should, of course, have available the entire range of applicable ASTM Standards.

Also refer to items marked [4] in the Bibliography.

Paint and Transparent Finishes

This chapter contains a discussion of paint and transparent finishes, as defined in the following paragraphs. It does not contain a discussion of coatings or special coatings, as defined here.

There is no generally accepted definition in the industry that clearly states the difference between paints and coatings. Paint manufacturers and other industry sources often use the terms interchangeably. The author has no quarrel with those sources on this matter. Nevertheless, because such dual use is confusing and makes discussion of the subject sometimes confusing, the following definitions apply to this book, if not to the industry as a whole.

Coatings. There are two categories of coatings: (1) coatings, and (2) special coatings. Coatings consist of materials that can only be applied in the shop or factory. They are used on metal, glass, porcelain, and other materials. They include such products as Kynar coatings on metal and anodizing on aluminum.

The special coating category includes relatively thick high-performance arthitectural coatings, such as high-build glaze coatings; fire-retardant coatings; industrial coatings, such as those used in sewage disposal plants; and cementitious coatings. Special coatings are usually applied in the field.

Paint. Paint is applied as a liquid in the field to form a thin opaque coating on the material to which it is applied. Primers for field-applied paints may be applied by hand or machine in the factory or shop, but are part of the paint system and are therefore also called paint.

In addition, all parts of a paint system are called paint, collectively and individually. Such parts include primers, emulsions, enamels, opaque stains, sealers, and fillers, and other applied materials used as prime, intermediate, or finish coats.

Transparent finish. A transparent finish is a system of materials that are applied as liquids by hand or machine in the field, factory, or shop to form a finish on wood through which the wood color, grain, or both are visible. The wood may appear as its natural color or its color may be darkened by the finish, lightened by a bleach, or altered by a stain. The grain may be enhanced, diminished, or concealed by the finish system. Components include bleaches, transparent stains, undercoats, and finish coats. The undercoats and finish coats are often lacquer, varnish, shellac, or polyurethane, but may be oil.

In discussing paint and transparent finishes, as defined above, the following generally accepted definitions apply.

Extender. A material used to impart some desirable quality to a paint that the paint alone does not have. Extenders may make a paint flow more easily, for example. They do not usually increase the hiding ability of the paint.

Film-former. Usually called a binder, a film-former is a nonvolatile ingredient in paint which binds the solid particles together.

Paint system. The several coats that are necessary to produce a complete paint coating are called collectively a paint system. Materials that are used as coats in paint systems are called paint. The various necessary coats include primers, emulsions, enamels, stains, sealers, fillers, and other applied materials used as prime, intermediate, or finish coats.

Pigment. The solid particles, usually a fine powder, in a paint or stain that provide the material's color and ability to cover and fill. Pigments are the ingredients that make paint opaque.

Solids. All the nonvolatile ingredients in a paint system's components, a stain, or a transparent coating are called solids. Solids include the binder (film former) and pigments.

Solvent. When used in defining a paint system component, stain, or transparent finish, a solvent is a volatile liquid that is used to dissolve the film former and pigment.

Stain. A material that is applied as a liquid in the field or in the factory or shop. Stains may be opaque or transparent. Stains may be part of a transparent finish or part of a paint system.

Thinner. A liquid used for thinning (reducing the viscosity of) a paint or transparent finish system component. When the component is solvent-based, the thinner is a volatile material. When the component is water-based, the thinner is usually water. The terms thinner and solvent are often used interchangeably.

Vehicle. The resins which form a flexible film and bind the pigment together are called the paint's vehicle. The vehicle contains the paint's film-former (binder).

Paint and Transparent Finish Materials

Characteristics Common to All Paint and Transparent Finish Materials

Manufacturers.   Paint and transparent finishes are manufactured and distributed by national, regional, and local manufacturers. At the time of this writing (1989) there are only nine manufacturers that can rightly be called national manufacturers of a broad range of paint and transparent finish products. They can be called national because they maintain service and distribution facilities in most major cities and regional locations. Their products are therefore easily and quickly available anywhere.

There are also other manufacturers that maintain service and distribution facilities nationally, but whose products are specialized in nature and limited in number.

There are also companies whose main business is to produce products for sale in retail outlets, such as Sears. Available nationally, their products are mostly used by individual consumers.

Regional manufacturers may operate production facilities as large as some of the national manufacturers, but have chosen to distribute their product mainly in only a portion of the country. In some cases, the regional manufacturers' products, which are often comparable to, and sometimes better than, the national manufacturers' products, are easier to obtain and result in a lower overall cost than do the national manufacturers' products.

There are hundreds of local manufacturers, whose products may be as good as those produced by national and regional manufacturers. In some cases, especially where a local area has climatic or other conditions that are different from those in most other areas of the country or even region, locally manufactured products will solve paint problems better than products produced by national or regional manufacturers. Local manufacturers may also be more readily available and eager to serve a local customer than some large national or regional manufacturer. Some building owners and managers prefer to deal with local businesses.

Specific manufacturers are not mentioned in this book. Refer to "Where to Get More Information" at the end of this chapter for more information about manufacturers.

Products. Paint and transparent finish products are available in several different quality levels. Some manufacturers make only what are called professional coatings, which are materials intended for use only by professional painters. Some make products that are available over the counter in various retail outlets. Some make products used only in cheap installations, such as the sprayed-on off-white coats used in new apartment buildings and the like. Some manufacturers make products in several or all three of those categories. Some make products that fit into various sublevels of those categories. The result is a bewildering array of products which is almost impossible for anyone who is not a paint professional to decipher.

To make the situation even worse, there are few recognized industry standards for paint or transparent finish products or systems. The only widely recognized standards for paint products are those contained in Federal Specifications. But the Federal Specifications are often out of date, inapplicable to a particular situation, and in some cases require products that no paint company offers as a standard product. Many paint manufacturers are capable of producing almost any paint formulation, but most will not do so. Especially formulated products will often be expensive compared to standard available products and will probably not have significantly superior characteristics.

There are hundreds of ASTM standards relative to paints, but most are for the component materials. Since any given paint product may contain a large number of different chemicals, using ASTM standards to control paint products is difficult and impractical.

The lack of standards does not mean that there are no legal restrictions. Environmental control laws may drastically alter paint and transparent finish material selection in areas where such laws are in force. Currently twenty-two states and the District of Columbia have enacted restrictive legislation regarding solvent emission. Some local jurisdictions have followed suite, in some cases passing even stricter legislation than the states. Such laws are expected to appear in other states and jurisdictions in the future.

To make a complicated situation even worse, paint industry members often do not speak the same language. As we alluded to earlier, a material that some call a paint, others call a coating. One company's best paint is another's finest. In some cases, even within a single company, the language used is confusing. Two qualities of paint may even have the same name.

Unfortunately, those of us who are not chemists or paint experts have little choice other than to locate and rely on reputable manufacturers and their reputable representative. Most manufacturers employ architectural or professional representatives, whose job it is to inform building owners and managers, architects, and contractors regarding their products. Some of them are based in regional offices, but many are available only at the manufacturer's national office. Read "Where to Get More Information" at the end of this chapter for additional discussion about the help that manufacturers furnish.

Those responsible for applying paint or transparent finishes should obtain, review, and understand the manufacturer's technical literature for each product to be used, including the label analysis and application instructions.

Paint system and transparent finish components should be delivered to the application site factory mixed, ready for application except for tinting and thinning, and in their original unopened containers, bearing the material name or title, manufacturer's name and label, and the standard to which it complies (Federal Specification number, for example), manufacturer's stock number, date of manufacture, contents by volume for major pigment and vehicle constituents, thinning instructions, application instructions, color name and number, and fire hazard data, where applicable.

Colors and Finishes.   A detailed color schedule should be made up and provided to the applicator. The color schedule should identify not only the major colors selected for each surface, but the color and location of accent colors, location of color changes (in a corner, for example), and colors for contrasting elements, such as trim, doors, and door frames.

Color pigments should be pure, nonfading, applicable types to suit the substrates and service required.

Covered and uncovered pipes, conduits, hangers, and ducts passing through a finished room or space are often painted with the type of undercoating materials consistent with the materials to be painted and finished with same type and color of finish coat as used on immediately adjacent wall or ceiling surfaces.

Prime-coated door hinges and butts and overhead closers are often painted the same color as the door trim to which they are attached. The top coats will probably be the same paint. Primers and undercoats may vary, of course, depending on the substrate materials.

Each coat in a paint system should be made a slightly different shade from the preceding coat. Final colors, glosses, and textures should match acceptable job-applied sample areas.

Paints

As mentioned earlier, the majority of paint products used in most buildings fall into one of two categories: (1) professional coatings, and (2) trade sales materials. The difference between the two has more to do with the formulation and method of handling and application than with quality.

Professional coatings (paint) products are formulated for application by professional painters who know how to mix and modify them to produce the best installation for the lowest price. They are available in large containers (even 55-gallon drums) as well as small. They usually produce a less costly installation than do trade sales materials.

It is a common misconception that paint materials sold across the counter (trade sales materials) in small containers are of lesser quality than those available directly from the manufacturer in large quantities. It is often, though not always, true that the need for packaging in smaller containers and for providing detailed labels increases the cost of trade sales materials and makes them slightly more expensive than professional products. It is not true that trade sales materials are of lesser quality than professional products. In fact, where the final installation is concerned, trade sales materials are easier to install and more or less foolproof, which may make the final installation superior. The final installation is certainly likely to be superior when anyone other than a professional experienced painter is applying the material. Many painting contractors, in fact, routinely use trade sales materials because of their ease of use and foolproof nature.

When properly mixed and applied, the final result will be equal in quality for comparable trade sales and professional products. There will be differences, however, in the composition of the materials that were applied. Ordinarily, the differences are of no significance, when the work was done properly and the composition was properly adjusted. Using the wrong materials to modify professional materials can sometimes have disastrous effects on performance, however, and modified professional materials may not be compatible with even the same material later applied but which has been modified using different chemicals. The field testing for compatibility mentioned later in this chapter thus becomes more important when the original paint was a professional one than when the more predictable trade sales materials were used.

The quality and expected service life of a paint are related to several characteristics, including color and gloss retention; adhesion; corrosion, mildew, and graffiti resistance; and hiding power (opacity). Another major quality determinant is the percentage of volume solids in the paint material that will remain on the painted surface after the paint has dried. The more volume solids there are, the thicker the dried film will be, and the greater its opacity. Conversely, the more water or solvent a paint contains, the lower its quality is likely to be. The composition of the solids does, however, have an effect on the paint's quality. If more latex solids or titanium dioxide are added, for example, the quality will improve.

Regardless of the type, paint should be well ground, should not settle badly, cake, or thicken in the container, should be readily broken with a paddle to a smooth consistency, and should have easy brushing qualities.

Exterior paint may be either chalking or nonchalking, but should be mildew-resistant.

Codes, regulations, or laws often control the acceptable lead content of paint used in residential applications and other locations where it will be accessible to children. Usually, paint for such applications should contain no more than 0.06 percent lead by weight calculated as lead metal in the total nonvolatile content of liquid paints or in the dried film of paint after application. More strict requirements may apply, however.

Paint Composition.   Paint is composed of four different groups of components. They are the vehicle; volatile or thinner, which are also called the solvent; pigment; and additives. Each group is composed of several different ingredients, and each serves a different function.

Vehicle.   A paint's vehicle carries the binder (film-former) to the surface to be covered and forms a film to bind the pigment particles together. It also gives the paint continuity and makes it adhere to the covered surface.

The type of binder in the vehicle is the characteristic by which most paints are identified. There are many binder types used in paints today. The more common are:

Latex. Paints with a latex film-former dry by the evaporation of water. Both interior and exterior paints are available with a latex binder. Latex paints are most often used on gypsum board surfaces, but are also used, with proper primers, on wood, concrete, masonry, plaster, and many other materials. They are the paint of choice in most current residential and many current commercial projects, and may be the most widely used paints today. The latex in latex paints is either polyvinyl acetate, polyacrylic, or polystyrene-butadiene. Latex paints go on easily and dry quickly, with little odor. There are many colors available and color retention is good.

Oil. Paints with an oil binder were used extensively in the past and are still used today, although in fewer cases. The oil in oil-based paint is usually, but not always, linseed oil. Most current use of oil-based paint is on the exterior of buildings, because they dry slowly. They are not recommended for use on masonry or concrete because of their sensitivity to alkalies. They provide high-quality applications over metals.

Alkyd. Probably the most common enamels used today have an alkyd base. When most people today refer to oil paint, they really mean alkyd, which are actually oil-modified resins. Alkyd paints are available that will produce either a flat, semigloss, or high-gloss finish. They are used over most materials, except directly on new concrete, masonry, or plaster. As are oil paints, alkyds are sensitive to alkalies. On most surfaces, particularly on wood and metal, alkyds produce a strong, long-lasting enamel finish.

Oil-alkyd combinations. When linseed oil is modifed with alkyd resins, a binder results that improves the paint by reducing its drying time, making it harder, and reducing its tendency to fade. Oil-alkyds are used as trim enamels for wood and metals, and as primers on structural steel.

Cement. Portland cement, lime, and pigments are the main ingredients of cement-based paints. Other components may include binders, additives, and sand. Cement-based paint is furnished in dry form for mixing with water at the application site. It is used primarily to coat rough concrete, masonry, or plaster as a base for a finish coat of paint. Cement-based paints are not usable in hot dry climates because they need wet conditions to promote curing. They do not block the flow of water vapor out of the coated material but may make the wall slightly less permeable to free water while permitting the wall to breathe.

Oleoresinous. Processing drying oils with hard resins produces an oleoresinous binder that is used as a varnish or mixing vehicle. They are seldom used today because they tend to turn yellow over time.

Phenolic. Products containing phenolic binders, an oleoresinous binder, are available as a pigmented paint or clear finish. The finish may be either flat or high gloss. Phenolics were among the first synthetic resins. They were used on exterior wood and on metals. They can be used in wet environments.

Rubber-based. While paints with latex binders are water-emulsion types, materials with rubber-based binders are solvent-thinned. The resins used are actually synthetic rubber. Rubber-based materials are lacquer-type paints that dry quickly to form water- and chemical-resistant surfaces in areas where water is a problem, such as in showers, laundry rooms, and kitchens. They can also be used on exterior surfaces. They are hard to recoat, however, because their strong solvents tend to lift the previously applied material.

Urethane. Material with an oil-free urethane binder is used as a pigmented material. Urethanes perform the same function as alkyds, but are more expensive and will not retain a gloss in exterior applications.

Epoxy-emulsion. Two-component epoxy-emulsion paints are used where a high-performance, low-odor paint is required.

Combinations. Some different types of binders can be combined to produce binders with characteristics often superior to either of the components. Some combinations include phenolic-alkyd, silicone-alkyd, and vinyl-alkyd.

Volatile or Thinner. Paints contain low-viscosity volatile solvents or other thinners that make them liquid and help them penetrate the surface to which they are applied. Volatile solvents and other thinners evaporate after the paint has been applied, and thus make up no part of the applied paint film.

In water-reducible paints, of course, the thinner is water.

Many types of solvents are used in solvent-based paints. Most alkyd paints contain alkyd solvents, or one or more petroleum-based organic material, such as mineral spirits, naphtha, xylene, or toluene. Other solvent-based paints contain other solvents such as ketone, acetates, alcohols, or glycol ethers.

In water-reducible paints, the thinner (water) separates the binder droplets. In solvent-based paints, the solvent dissolves the binder and holds it in solution. Solvents and water thinners also keep the paint at the proper viscosity and control its setting time. The proper amount of solvent or water, as appropriate, is essential to proper paint performance. That is why proper thinning of paints by the applicator is essential. It is also the reason that many trade-sales products require application without thinning.

It is the volatile solvents in paints and their necessary evaporation that produce most of the pollution that paint contributes to the atmosphere. Currently used solvents are either hydrocarbon or oxygenated types. They both contain hydrogen and carbon atoms, but the oxygenated type also contains oxygen. The hydrocarbon type is cheaper but runs afoul of current laws in some locations and may eventually give way entirely to the oxygenated type. Ultimately, both types may disappear, but their replacements are not yet available.

Older paints used terpene solvents containing such materials as turpentine or pine oil, but they are seldom used today.

Pigments.   The solid material that is left when a paint dries is mostly pigment. Pigment particles are permanently insoluble. They dictate the paint's color, opacity, and gloss, and resist corrosion, weathering, and abrasion of the paint's surface. They also contribute to the paint's hardness and govern its adhesion to the surface.

A paint's gloss is determined by its pigment-to-binder ratio. Larger pigment particles produce less gloss. Extender pigments, such as clay, calcium carbonate, or silica may also be used to reduce gloss. Gloss is rated as flat, eggshell, semigloss, and high gloss. Usually, opaque materials with a sheen are called enamel, but flat enamels are also available today.

Opacity is a paint's ability to cover the surface and hide what lies below the paint. Opacity is a function of the pigment.

Additives.   Many materials are added to paints to alter or improve their basic characteristics. Some are essential. Additives include plasticizers, air dryers, oxidizing catalysts, converters, fungicides, preservatives, and wetting, antisetting, and antiskinning agents.

Paint Systems.   A single paint coat used alone will be adequate only under limited circumstances. A single coat might be sufficient to produce an initial surface on gypsum board in a new condominium building or tenant space in an office building, for example, where it is expected to be only a base coat for materials that will be added later by the tenant. In most circumstances, a system of materials must be used in combination to form a satisfactory paint application. All coats in a paint system must be compatible for a successful installation. Barrier coats should be applied over incompatible primers or the incompatible primers should be removed and a new primer applied.

Most paint manufacturers routinely furnish specific recommendations for the several components in the paint system that should be used for most substrates and conditions. Paint systems used should be in accord with those recommendations. In addition, each paint system proposed for use should be reviewed by the manufacturers of the products to be used to ensure compatibility of the various coats of the system with each other and with the materials they will contact, and the suitability of the system for the conditions. When the manufacturer(s) suggests that different products or combinations of products be used, other than those selected (a different prime coat, for example), its suggestions should be followed.

Paint systems recommended by a paint manufacturer should be checked against the advice of the manufacturer of the material to be painted and the association representing that manufacturer, to verify their suitability for the substrate and conditions. The Portland Cement Association, National Concrete Masonry Association, Brick Institute of America, and similar organizations offer specific advice regarding the proper paint to use on the products they represent. Their advice should be heeded. In most cases, a reading of their recommendations, as contained in the publications referenced in this book, is adequate to verify a paint manufacturer's advice. When there is any doubt or disagreement, the applicable association should be contacted directly.

Most paint systems consist of a primer and a top coat. In some systems, an intermediate coat is also used. Intermediate coats are usually applied in systems with a glossy sheen. A block filler is usually used over porous unit masonry.

Unless there is some compelling reason to do otherwise, all the products used in a single paint system should be produced by the same manufacturer or, at least, be supplied by the same company that manufactures the finish coats. Thinners should be approved by the manufacturer of the material being thinned. Undercoats and primers, whether shop- or field-applied, should be approved by the finish coat manufacturer.

Primers are paints that have been formulated for application to bare surfaces. They serve as a base for succeeding coats and have to be formulated to adhere readily to the substrate and the succeeding coat. They must also be compatible with both the substrate and the top coats. Oil-based and alkyd primers should not be used on galvanized surfaces, for example, because their presence can create a chemical reaction that leaves a soap on the metal's surface, which will not permit the paint to adhere. Some primers, such as rust-inhibitive primers, also serve the major secondary function of protecting the substrate. Not all substrates require special primers with all paints. Latex paints may be applied directly to gypsum board, for example. Porous substrates and substrates containing alkalies may require special primers called primer-sealers designed to ensure coverage or to protect the succeeding coats from the substrates. Some primers serve to isolate essentially incompatible paint and substrates. Primers on galvanized metal surfaces, for example, separate the galvanizing from the top coats, many of which contain chemicals that can react with the zinc coating on the metal.

Enamel undercoaters are used beneath enamel top coats.

Fillers are used to provide a smooth surface and level out rough surfaces of unit masonry.

Top coats are the main barrier to weather, chemicals, soiling, and abuse. They also protect the undercoats and primers, and provide the finished surface.

Special Paints.   Paints specifically formulated to handle unusual conditions are available. For example, paints that are intended for use on cleaned metal are generally called direct-to-metal paints. There are, however, also direct-to-rust paints which are epoxy mastic materials formulated for direct application over rust. There are also rust conversion coatings that convert ferric oxide (rust) to a stable organic iron compound, which then becomes part of the coating.

Textured Surfaces.   Textured surfaces may be produced in many ways. Some materials can be given a texture during manufacture or production. The surface of cementitious materials, such as plaster, may be altered to produce a textured finish. Integral and factory-applied finishes are beyond the scope of this book.

Products called special coatings are also used to produce textured surfaces. There are many formulations, ranging from solvent-based materials to water-based acrylic emulsions. Special coatings are beyond the scope of this book.

Early textured paint finishes were produced by adding sand to the paint. Many such finishes still exist, but most paint-like textured finishes today are produced using products made specifically for that purpose from materials similar to those used in gypsum board joint compounds. These products are manufactured by the same companies that make gypsum board products. Earlier versions, which became very popular in the 1920s, were called plastic paints. Many of the early products contained asbestos. So-called plastic paints and their successors are beyond the scope of this book.

Stains.   Interior stains used as a part of a transparent finish are discussed under "Transparent Finishes." Wood preservatives, which might be clear, semi-transparent, or opaque are beyond the scope of this book.

There are also several types of wood stains designed for exterior use. Semi-transparent stains are oil-based materials that dye the wood fibers as they penetrate. They cannot be used over previously painted or sealed surfaces, and they must be renewed frequently, perhaps as often as once a year, since they do not protect for long. Pigmented semi-transparent exterior wood stains, especially in the dark colors, tend to hide the natural bleeding that occurs when they begin to fail, and therefore may need renewal somewhat less frequently.

The other types of exterior wood stains include semi-solid stains, oil-based solid color stains, acrylic solid color stains, weathering stains, bleaching stains, deck stains, and others. Most of them are weather resistant. Many contain water repellents and wood preservatives. They usually require re-coating less frequently than do semi-transparent stains, but tend to fade fairly rapidly. Oil-based stains may require recoating within a year or two. Wood stains containing 100 percent acrylic latex tend to last longer and fade slower than oil-based stains, however, and may go virtually unattended for three years or longer.

No exterior wood stain will last as long or resist fading as well as paint will.

Transparent Finishes

The transparent finishes discussed in this chapter include oiled, stained, and natural finishes for wood. Transparent finishes for metals are beyond the scope of this book.

Transparent Finishing Materials.   Transparent finishes are achieved using the following materials, either alone or in combination.

Wood Stain.   Stains used to color wood in interior applications are made from color pigments suspended in linseed oil, or another drying oil. Stains are then thinned to make them easy to apply. Stained wood may be finished with linseed oil or one of the clear finishes discussed in the following pages.

Linseed Oil.   Boiled linseed oil constitutes the classic oil finish. It is inexpensive, easy to apply, maintain, and repair. Normal applications require reducing the oil by mixing it with turpentine in ratios ranging from equal parts of each to twice as much oil as turpentine.

Varnish.   A liquid in the can, varnish is a homogenous mixture of resin, drying oil, drier, and solvent. When varnish dries, it forms a transparent or translucent film, which may be flat, satin, or high gloss. It is available in colors ranging from nearly clear to dark brown, and in several qualities. The higher-quality products expand and contract without cracking. Varnish is classified as short-oil, medium-oil, or long-oil, depending on the number of gallons of oil that it contains for each one hundred pounds of resin.

Other materials are added to some varnish to impart qualities other than those that are natural to the material. Pigments, such as synthetic silica, are added to change the natural high-gloss sheen of varnish to a low-gloss finish. When additives are included to give the material a resistance to salt water, it is called spar varnish.

Varnish is used as a clear finish and, when reduced by solvents, as a sealer for wood and plywood. Varnishes for interior use usually include alkyd resins, but epoxy-ester varnishes are also available.

Many varnishes formulated for exterior use are made from tung oil and phenolic resin, although some other formulations are also used. Sometimes, varnishes intended for interior use are also used on the exterior and may function satisfactorily if well protected from the weather.

Shellac.   Shellac produces a similar finish to that produced by varnish. It is available in either white or orange. White shellac can be tinted with alcohol-soluble aniline dyes, which makes it especially valuable for blending repairs with old work. Orange shellac produces a finish with a deeper tone than does the undyed white material. Shellac has a relatively short shelf-life, being often unusable within four to six months after manufacture.

Lacquer.   Although available in formulations which will produce either flat or glossy finishes, lacquer is seldom used for field-applied finishing of building surfaces. It is more likely to be used for furniture or casework finishing in the factory or shop. Lacquers have the distinct disadvantage of not being usable over existing finishes. They can actually be used as paint removers.

Polyurethane.   Some references and some manufacturers classify poly-urethane clear finishes as varnish. Others call them lacquers. In fact, though they have some characteristics of each, they are neither. Urethane finishes are available as either oil-modified or moisture-curing types.

Oil-modified polyurethanes are clear materials only. They are sometimes used in exterior applications, but are better suited to interior locations. Urethanes are water-resistant and highly durable, which makes them usable on floors and in such wet locations as counter or bar tops. Clear urethanes are available in formulations that will produce either gloss or matte finishes. They tend to not hold a gloss, however, when used on the exterior. Some sources say that they do not bond well to existing finishes, others say that they may be so used. They cannot be used over shellac, paste wood fillers, and some other finishes. The individual manufacturer's recommendations should be followed when determining where a urethane finish can be used.

Moisture-curing polyurethanes may be clear or pigmented to produce a colored opaque finish. Since they are dependent on water to cure, they do not work well in very dry climates.

Wood Sealers.   Penetrating wood sealers are actually wood preservatives and not transparent finishes. They are generally beyond the scope of this book. They may create finishes that resemble some transparent finishes, and the evidence of their failure may resemble those discussed in this chapter for transparent finishes. The expected life span of many such materials is one year or less. When they are not renewed within that interval, the protected wood may show evidence of water and weather damage, such as bleeding, mildew growth, and general blackening. The manufacturer of such products should be consulted and its advice followed.

Transparent Finish System.   A single coat of a transparent finish material used alone will almost never be adequate. Usually, a system of materials must be applied to form a satisfactory transparent finish. The number of coats and the composition of the coats depends on the desired effect. Generally, at least three coats of the finish material are needed to provide an acceptable level of durability and appearance, though some of those coats may be cut with the proper thinning agent. Where a color different than that natural to the material being covered is desired, adding a stain coat is necessary. A wood filler is usually used to provide a smooth surface on open-grain woods.

All coats in a transparent finish system must be compatible for a successful installation. Unless there is some compelling reason to do otherwise, all of the products used in a single transparent finish system should be produced by the same manufacturer, or at least be supplied by the same company that manufactures the finish coats. Thinners should be approved by the manufacturer of the material being thinned.

Most manufacturers of transparent finish materials routinely furnish specific recommendations for the several components in the transparent finish system, which should be used for most substrates and conditions. Systems used should be in accord with those recommendations. In addition, each transparent finish system proposed for use should be reviewed by the manufacturers of the products to be used to ensure the compatibility of the various coats of the system with each other and with the materials they will contact, and the suitability of the system for the conditions. When the manufacturer(s) suggests that different products or combinations of products be used, other than those selected, its suggestions should be followed.

Miscellaneous Materials

Many related materials are necessary to produce effective paint and transparent finish systems. They include crack and seam fillers for wood, concrete, and mortar; wood filler; plastic wood; turpentine; linseed oil; mineral spirits; denatured alcohol; lacquer thinner; and others. For standards applicable for those and other materials needed for paint and transparent systems applications, refer to "Where to Get More Information" at the end of this chapter.

Paint and Transparent Finish Applications

Requirements Common to All Paint and Transparent Finish Applications

Surfaces That Are Usually Painted. Almost any new surface may be painted, but painting and transparent finishes are usually applied on the following types of surfaces.

Exterior exposed steel and iron, including miscellaneous metals, hollow metal doors and frames, lintels, roof-mounted equipment supports, mechanical and electrical equipment and devices, and similar items.

Exterior galvanized steel and non-anodized aluminum, including miscellaneous metals, louvered penthouses, mechanical and electrical devices and equipment, and the like.

Roof-mounted mechanical and electrical equipment and devices, and roof accessories, including ventilators, except that factory finished surfaces are not usually field-painted.

Exterior copper, lead-coated copper, stainless steel and other flashings, and other exterior sheet metal exposed to view.

Sectional overhead doors.

Exterior gypsum board.

Walls and ceilings of interior spaces, excluding parking areas, but including stair towers, mechanical, electrical, and telephone spaces.

Parking markings.

Interior exposed miscellaneous steel and iron, galvanized steel, structural steel, hollow metal work, permanent concrete formwork, steel stairs, specialty items, and the like.

Wood shelves, standing and running trim, doors, windows, window trim and stools, plywood, and other exposed interior wood not otherwise finished. Wood doors, cabinetwork, and other hardwood items are often given a transparent finish. Such items may be field- or shop-finished.

Shop-applied coats on miscellaneous steel and iron, and primed items of equipment, often require touch-up in the field.

Piping, conduit, ducts, insulation coverings, motors, pumps, enclosures, machinery, and the like that are exposed to view in usable and habitable spaces, including mechanical, electrical, telephone, and storage rooms are usually painted. Galvanized and aluminum items and duct coverings in mechanical, electrical, telephone, and storage rooms, and surfaces which are factory finished, are often left unpainted.

Heating, ventilating, and air conditioning cabinets, equipment, and devices that were given a prime coat in the factory.

Electrical equipment and devices that were given a prime coat at the factory. Such items include panels, trim, circuit boards, switches, and the like.

Factory-primed surfaces, such as hollow metal doors and frames, elevator hoistway doors and frames, coiling doors, fire extinguisher cabinets, and access doors and frames.

Surfaces That Are Usually Not Painted. The following types of surfaces are usually not field painted in new construction.

Surfaces that are concealed and generally inaccessible, such as inside furred areas, chases, spaces above ceilings, utility tunnels, crawl spaces, and shafts, and piping, equipment, and other items occurring in those spaces.

Exterior cast-in-place concrete of all types, including traffic surfaces.

Concrete floors, stair treads, and stair risers. Of the items listed here, concrete floors are the most likely to be painted.

Code-required labels, such as Underwriters Laboratories and Factory Mutual, or any equipment identification, performance rating, name, or nomenclature plates.

Operating parts such as motor and fan shafts, linkages, valve and damper operators, and sensing devices.

Materials and items with a factory or integral finish. The materials and items in the following list often, although not always, have a factory-applied or integral finish that requires no field-applied finish. The list is by no means complete.

- Pavers, including concrete unit pavers, paver brick, and most other pavers
- Cast-in-place concrete
- Precast concrete
- Natural and cast stone
- Face brick and brick flooring
- Anodized aluminum
- Stainless steel
- Chromium plate
- Copper
- Brass
- Bronze

- Baked enamel
- Porcelain enamel
- Fluorocarbon finishes
- Laminated wood shapes
- Laminated plastic assemblies
- Plastic assemblies
- Slate, tile, concrete, and metal roofing tiles
- Aluminum entrances, storefronts, sliding doors, sloped glazing, wall louvers, skylights, and windows
- Curtain wall components, including panels
- Ceramic and quarry tile
- Terrazzo
- Acoustical ceilings, wall panels, and baffles
- Resilient flooring; bases; stair risers, treads, and stringers
- Carpet
- Resinous, epoxy, and other liquid-applied flooring
- Special coatings, including high-build glazed coatings, cementitious coatings, and others
- Wall coverings, including vinyl wall covering, fabrics, wallpaper, and the like
- Chalkboards and tackboards
- Toilet and entry partitions and doors, shower enclosures, urinal screens, dressing compartments, and cubicles
- Wall and corner guards and crash rails
- Fireplace accessories
- Flagpoles
- Signs, plaques, and other identifying devices
- Lockers
- Folding and other operable gates, doors, and partitions
- Demountable partitions
- Metal storage shelving and associated units
- Toilet and bath accessories
- Shower rods
- Curtain tracks
- Equipment items of all sorts, including equipment for use in laboratories, libraries, darkrooms, athletic and recreational facilities, food preparation and serving areas, parking control, audio-visual applications,

retail stores, coatrooms, and many other uses. This category includes unit kitchens and residential appliances

- Manufactured casework for all uses, ranging from residential kitchens to medical facilities
- Audiometric, athletic, saunas, and other special-purpose prefabricated rooms
- Pre-engineered metal structures
- Pools
- Solar energy systems
- Dumbwaiter and elevator hoistway entrances and cars
- Escalators and moving walks
- Lifts
- Material handling systems
- Heating, ventilating, and air conditioning equipment and devices
- Plumbing fixtures
- Shower receptors and stalls
- Janitor's receptors and sinks
- Electrical equipment and devices

Preparation for Painting and Applying Transparent Finishes

General Requirements. Before general painting or finishing begins, a complete wall, space, or item selected for each principal color, should be painted or finished for each principal paint and transparent finish that will be used. The sample should show color, finished degree of gloss and texture, materials, and workmanship. The sample should be repainted or refinished until acceptable. The acceptable sample wall, space, or item should then serve as the standard for similar work.

Paint and transparent finish materials should be stored in a single place for each major portion of the work. Storage places should be kept clean. Paint and transparent finish materials should not be stored in closets or other small confined locations.

Oily rags, oil, and solvent-soaked waste should be removed from the buildings at the end of each work day. Precautions should be taken to avoid the danger of fire.

Water-based paints should be applied only when the temperature of the surfaces to be painted and the surrounding air is between 50 and 90 degrees Fahrenheit, unless the paint manufacturer's printed instructions say otherwise.

Solvent-thinned paints and transparent finish materials should be applied only when the temperature of the surfaces to be painted and the surrounding air is between 45 (some sources say 50) and 95 (some sources say 120) degrees Fahrenheit, unless the paint manufacturer's printed instructions recommend otherwise.

Paint and transparent finishes should not be applied in snow, rain, fog, or mist; or when the relative humidity exceeds 85 (some sources say 90) percent; or to damp or wet surfaces; or to extremely hot or cold metal (Fig. 5-1), unless the material manufacturer's printed instructions recommend otherwise. Cement-based paint, for example, should be applied over a surface that has been dampened by a water spray. Painting of surfaces exposed to the hot sun should be avoided. Painting may be continued during inclement weather, however, if the areas and surfaces to be painted are enclosed and heated to within the temperature limits specified by the paint manufacturer during the application and drying periods.

Once painting has been started within a building, a temperature of 65 degrees Fahrenheit or higher should be provided in the area where the work is being done. Wide variations in temperatures, which might result in condensation on freshly painted surfaces, should be avoided.

Figure 5-1 This photo shows the results of painting hot metal. Most of the blisters have broken and the paint has flaked off. (*Photo by author.*)

Inspection. Areas to be painted or to receive a transparent finish, and conditions under which paint or a transparent finish is to be applied, should be inspected carefully. It is not unusual to find conditions other than those that are expected. Unsatisfactory conditions should be corrected before work related to the paint or transparent finish is begun.

When surfaces to be painted or finished cannot be put in proper condition by customary cleaning, sanding, and puttying, do not proceed until proper conditions have been achieved using other accepted methods.

Each coat of a paint or finish system should be inspected and found to be satisfactory before the next coat is applied.

General Requirements for Surface Preparation. Surface preparation is certainly among the most important aspects of a paint application. It will often determine whether the application will be a successful one. Some sources indicate that as high as 80 percent of paint failures are caused by improper surface preparation. Preparation and cleaning procedures should be done in accordance with the paint or transparent finish material manufacturer's instructions for each specific substrate condition.

Hardware, accessories, machined surfaces, plates, fixtures, and similar items not to be painted or finished should be removed before surface preparation for painting or finishing begins. Alternatively, such items may be protected by surface-applied tape or another type of protection. Even when such items are to be painted, they are often removed to make painting them and the adjacent surfaces easier. After the painting or finishing in each space or area has been completed, the removed items can be reinstalled.

Surfaces should be cleaned before the first paint or transparent finish coat is applied and, if necessary, also between coats, to remove oil, grease, dirt, dust, and other contaminants. Activities should be scheduled so that contaminants will not fall onto wet, newly painted or finished surfaces. Paint or finish should not be applied over dirt, rust, scale, oil, grease, moisture, scuffed surfaces, or other conditions detrimental to formation of a durable paint or finish film.

Mildew should be removed and neutralized by scrubbing the affected surfaces thoroughly with a solution made by adding two ounces of trisodium phosphate-type cleaner and eight ounces of sodium hypochlorite (Chlorox) to one gallon of warm water. A scouring powder may be used if necessary to remove mildew spores, but care must be taken to prevent damage to the surface being cleaned. Cleaned surfaces should be rinsed with clear water and allowed to dry thoroughly.

Each water-stained interior surface and the interior surface of exterior walls to which paint is to be applied directly (no supported finish, such as plaster or gypsum board) should first be primed. The primer should be an oil-based paint product specifically recommended by its manufacturer and the finish coat manufacturer to conceal water stain marks, so that they will not show through the finish paint coats. This prime coat should be in addition to the number of coats recommended by the paint manufacturer as required to produce the paint system to be applied on the affected surface.

Before painting or transparent finishing is started in an area, that space should be swept clean with brooms and excessive dust should be removed. After painting or finishing operations have been started in a given area, that area should not be swept with brooms. Necessary cleaning should then be done using commercial vacuum cleaning equipment.

Surfaces to be painted or finished should be kept clean, dry, smooth, and free from dust and foreign matter that would adversely affect paint adhesion or appearance.

Preparing Concrete, Stone, Concrete Masonry Units, and Other Masonry.

Concrete, stone, concrete masonry units, and other masonry to be painted should be prepared to receive paint by removing efflorescence, laitance, chalk, dust, dirt, grease, oils, and stains, and by roughening as required to remove glaze.

Materials used for cleaning may leave a residue that can itself reduce or prevent paint bond. Acids, caustic sodas, trisodium phosphate, detergents, and other such materials must be completely neutralized and removed before the surfaces are painted.

The methods and materials used for cleaning should not harm the surfaces being cleaned or adjacent surfaces. Abrasive blast-cleaning methods may erode surfaces, and should be used only where recommended by the paint manufacturer and the producer, or the association representing the producer, of the substrate being cleaned. Acids are dangerous, and should be used sparingly and with caution. They should not be used at all on concrete masonry units. Sometimes, however, when used by trained professionals, acids offer the best solution to a concrete preparation problem. Concrete floors to be painted, for example, should be washed with an etching cleaner, such as a 5 percent solution of muriatic acid, flushed with clear water to remove the acid, neutralized with ammonia, and permitted to dry completely before paint is applied. Efflorescence that does not respond to wire brushing, where sandblasting is not appropriate, might be removed from concrete using a 5 or 10 percent solution of hydrochloric acid.

Concrete that has a smooth surface imparted by the formwork to which paint will not readily adhere should be roughened by waterblasting, light sandblasting, or grinding with silicone carbon stones.

The alkalinity and moisture content of surfaces to be painted should be ascertained using appropriate tests. When the surfaces are sufficiently alkaline to blister or burn the paint, the condition must be corrected before paint is applied. Paint should not be applied where the moisture content exceeds that permitted in the paint manufacturer's printed directions. The Portland Cement Association (PCA) says that concrete should be completely dry.

Concrete and masonry surfaces should be permitted to cure completely, which usually takes 60 to 90 days even under moderate weather conditions. PCA says that paint should not be applied until concrete is well cured for 28 days to 6 months. The actual time, they say, should depend on surface conditions and the type of paint. It is generally agreed in the industry that paint should not be applied to such surfaces until they are at least 28 days old under any circumstances. Even masonry paint, which requires water to set properly, should not be applied earlier than that. The National Concrete Masonry Association recommends against applying oil-based or alkyd paints to concrete unit masonry that is less than six months old, unless the masonry is first treated with a 3 percent phosphoric acid solution and then with a 1 or 2 percent zinc chloride solution.

The exact type of form release agent used on concrete should be determined and removed using means appropriate to the type used. The form release agent and paint manufacturer's printed instructions for removal should be strictly followed.

Curing membranes should not be used on concrete to be painted. When curing membranes have been used, they must be removed completely or the paint will not bond. Removing such substances requires sandblasting, waterblasting, scarifying, or some other abrasive process.

Concrete should be painted with alkali-resistant materials. When paints that are not alkali-resistant, such as most alkyd- or oil-based paint, will be used, the PCA recommends pretreating the concrete with a 3 percent phosphoric acid solution and a 2 percent zinc chloride solution. There are some limitations, however. In such a case, it is best to request advice from the PCA.

Concrete surfaces to be painted should have pits, holes, and voids filled, as recommended by the PCA. Pits and depressions should be patched. Large holes should be filled.

Painting of unit masonry should not be started until tooling and pointing of the mortar joints have been completed and the mortar is thoroughly dry.

The PCA publications listed in "Where to Get More Information" near the end of this chapter contain specific recommendations for preparing concrete to receive paint.

Preparing Brick Masonry.   Brick is not generally intended to be painted, and new brick seldom is. Brick materials and construction should be the same quality, whether the brick will be painted or not. Lesser materials or construction should not be imposed in the vain hope that a coat of paint will conceal imperfections or produce a better brick construction. Of particular importance is the prevention of efflorescence.

Brick to be painted should be prepared to receive paint by removing efflorescence, chalk, dust, dirt, grease, oils, and stains. The methods used should not harm the surfaces being cleaned or adjacent surfaces. Acid cleaning solutions should not be used on brick that will be painted.

The alkalinity and moisture content of surfaces to be painted should be ascertained using appropriate tests. When the surfaces are sufficiently alkaline to blister or burn the paint, the condition must be corrected and the alkalinity neutralized before paint is applied. Zinc chloride and zinc sulfate are often used as neutralizers. Then, an alkali-resistant primer should be used. Paint should not be applied where the moisture content exceeds that permitted in the paint manufacturer's printed directions or in the Brick Institute of America's *Technical Notes on Brick Construction* "6: Painting Brick Masonry." Brick surfaces should be permitted to cure completely, which may take from 60 to 90 days, even under moderate weather conditions. Paint should not be applied to such surfaces until they are at least 30 days old. Masonry-paint, which requires water to set properly, may be applied, however, after the substrate has cured for about one month.

Painting of masonry should not be started until tooling and pointing of the mortar joints have been completed and the mortar is thoroughly dry.

The recommendations of the Brick Institute of America, as contained in the publications listed in "Where to Get More Information" later in this chapter, should be strictly followed when preparing for or painting brick masonry.

Preparing Wood.   Surfaces to be painted or finished should be cleaned of all dirt, oil, and other foreign substances using scrapers, mineral spirits, and sandpaper, as required (Fig. 5-2). Wood surfaces and edges should be sanded smooth before finishing or painting is begun, and between coats. Dust should be removed after sanding. Residue should be removed from knots, pitch streaks, cracks, open joints, and sappy spots. On wood surfaces to be painted, a thin coat of white shellac should be applied to pitch and resinous sapwood before the prime coat is applied. Small, dry, seasoned knots should be scraped and cleaned and a thin coat of white shellac or other recommended knot sealer should be applied over them before the prime coat is applied. After the primer has been applied, holes and imperfections in the finish surfaces should be filled with putty or plastic wood-filler. The surfaces should be sandpapered smooth when they have dried.

New wood to be painted or finished should be primed, stained if appropriate, or sealed immediately upon delivery. Edges, ends, faces, undersides, and backsides, of trim, cabinets, counters, paneling, and other wood should

Figure 5-2 Paint on this wood window sill failed because the previous paint was not properly prepared. (*Photo by author.*)

be primed. The primer should be an enamel undercoat, penetrating sealer, or varnish of the same material as will be used for the first coat of the exposed finish. Wood should never be primed, either in or on a building, during masonry erection.

Nails fastening wood to be painted or finished should be set and such screws should be countersunk. After the prime coat has been applied and has dried, the nail and screw holes and cracks, open joints, and other defects should be filled with putty or wood filler. Where a transparent finish will be used, the filler should be tinted to match the color of the wood.

Preparing Metals. Ferrous metal surfaces, which are not galvanized or shop-coated, should be cleaned of oil, grease, dirt, loose mill scale, and other foreign substances, using solvent or mechanical cleaning methods as recommended by the paint manufacturer. The recommendations of the Steel Structures Painting Council should be followed, unless they conflict with the paint manufacturer's recommendations. Bare and sandblasted or pickled ferrous metal should be given a metal treatment wash before the primer is applied.

It is necessary to remove dirt, oil, grease, defective paint, and rust from shop-coated ferrous, galvanized, and nonferrous metal surfaces, using a combination of scraping, sanding, wire brushing, sand blasting, and cleaning agents recommended by the paint manufacturer. Surfaces should be wiped clean before they are painted. Shoulders along the edges of existing paint should be sanded down when necessary to prevent them from telegraphing through the new paint. Where necessary to produce a smooth finish, existing paint should be completely stripped to the bare metal.

Successfully painting galvanized metal is one of the more difficult problems related to paints. One reason is that authoritative sources disagree about the proper methods and materials that should be used to prepare galvanized metals to receive paint. For example, some sources say that galvanized metal should be allowed to weather at least six months before it is painted. Other sources say that weathering is a bad idea because galvanized metal weathers unevenly, which can cause poor paint adhesion; unless the metal is directly exposed, it will not weather appreciably anyway; and even more preparation is required when the metal has been allowed to weather, and the weathering will make the preparation more difficult to accomplish properly.

Since unprotected galvanized metal will form "white rust," fabricators often coat galvanized surfaces with oils, waxes, silicons, or silicates. Those products must be removed before paint is applied. Some sources say that bare galvanized surfaces should be cleaned free of oil and surface contaminants with mineral spirits or xylol. Others say to use a solvent wash, and specifically recommend against using mineral spirits. Some say use petroleum spirits. Others say not to use petroleum-based solvents.

If white rust has formed, it must be removed by wirebrushing, sanding, or blasting. Normal rust which occurs where the zinc coating has been damaged should also be removed.

After all oils, white rust, and other containments have been removed, some sources say that the bare galvanized surfaces should be pretreated. Some sources say to use a diluted vinegar solution, which is a weak acetic acid. Others say to use a proprietary acid-bound resinous or crystalline zinc phosphate preparation, or phosphoric acid. Still other sources recommend that neither vinegar nor acetic acid be used on galvanized metal.

The best solution to determining the proper preparation methods to use on galvanized metals is to ask the paint manufacturer for recommendations and follow them. At least then there will be someone to complain to if paint failure occurs.

Bare aluminum should be washed with mineral spirits or turpentine, and allowed to weather one month or be roughened with steel wool.

Bare stainless steel should be washed with a solvent and prepared according to the paint material manufacturer's recommendations.

Bare copper should have stains and mill scale removed by sanding.

Then dirt, grease, and oil should be removed using mineral spirits. If a white alkyd primer was used, the copper should be wiped clean using a clean rag that has been wetted with Varsol.

Bare lead-coated copper should be sanded to bright metal. Grease and oil should be removed using a solvent recommended by the metal manufacturer.

Bare terne-coated metal should be cleaned with mineral spirits and wiped dry with clean cloths. Immediately after cleaning, the surfaces should be given a coat of a pretreatment specifically formulated for use on terne-coated metal. The pretreatments should contain 95 percent linseed oil. The pretreatment should be allowed to dry for at least 72 hours, and longer if necessary to ensure that it is thoroughly dry, before the primer is applied.

Welds, abrasions, factory- or shop-applied prime coats, and shop-painted surfaces should be touched-up after erection and before caulking and application of the first field coat of paint. The touch-up material should be the same as the primer. Touch up surfaces which are to be covered, before they are concealed. Touch-ups should be sanded smooth.

Preparing Gypsum Board.   General surface preparation of gypsum board should be done by the installer in accordance with the standards of that trade. Remaining cracks, depressions, holes, and other irregularities should be filled with spackling compound. Rough or high spots left by joint cement or spackling should be sanded smooth. Care should be taken to ensure that the gypsum board's surface is not damaged. Small damage that does occur and other surface defects should be repaired smooth.

Joints, cement-filled dimples, and patches should be spot primed and allowed to dry thoroughly before the primer is applied over the remaining surface.

Preparing Plaster, Stucco, and Mineral-Fiber-Reinforced Cement Panels.   Figure 5-3 shows the result of improperly preparing stucco for painting.

Stucco and other cement plaster should be allowed to cure for at least 60 days before it is painted. Other full-coat plaster, including patches, should be allowed to dry thoroughly for at least 30 days. Veneer plaster and skim-coat plaster should be allowed to dry thoroughly for the time recommended by the plaster manufacturer, but not less than 24 hours.

After the stucco and plaster patching ordinarily done by the plasterer has been completed, remaining scratches, cracks, and abrasions in stucco or plaster surfaces, and openings adjoining trim, should be cut out and filled with prepared patching plaster flush with the adjoining stucco or plaster surface.

Figure 5-3   Paint deterioration on ill-prepared stucco. (*Photo by author.*)

Minor imperfections in veneer plaster and skim-coat plaster should be repaired using spackling compound or joint compound recommended by the plaster manufacturer for the purpose.

Imperfections in mineral-fiber-reinforced cement panels should be repaired in accordance with the manufacturer's instructions. The assumption is here made that new panels are asbestos-free. Should panels containing asbestos be used, all precautions associated with such use should be strictly observed.

When the patching is dry, patched areas should be spot-primed before the overall prime coat is applied. When the spot priming has dried, the entire surface should then be covered with the prime coat that is part of the paint system, as recommended by the paint manufacturer.

Free lime should be neutralized by applying a solution of 2-1/2 pounds of zinc sulfate per gallon of water and allowing it to dry thoroughly before applying the first coat. Suction spots or "hot spots" should be touched-up after application of the first coat, to produce an even finish.

Before painting starts, surfaces to be painted should be tested for dryness with a moisture-testing device designed for the purpose. Paint or sealer should not be applied when the moisture content of the surface to receive the paint exceeds 10 percent, as determined by the testing device, except as may otherwise be recommended by the manufacturer of the paint materials to be used.

Sandpaper should not be used on plaster surfaces.

Preparing Other Existing Surfaces.   Surfaces not mentioned in this chapter should be prepared for repainting or refinishing in accordance with the paint or finishing material manufacturer's recommendations.

Applying Paint and Transparent Finishes

General Application Requirements.   The type of material for each coat, location, and use should be as designated for the purpose by the paint or finishing material manufacturer, as stated in manufacturer's published specifications.

Painting and finishing materials should be mixed, thinned, and applied in accordance with manufacturer's latest published directions. Materials should not be thinned unless the manufacturer specifically so recommends. Applicators and techniques should be those best suited for the substrate and the type of material being applied.

Sealers and undercoats should not be varied from those recommended by the paint or finish manufacturer.

Materials not in actual use should be stored in tightly covered containers. Containers used in the storage, mixing, and application of paint should be maintained in a clean condition, free of foreign materials and residue.

Materials should be stirred before application to produce a mixture of uniform density, and stirred as required during application. Surface films should not be stirred into the material. Film should be removed and, if necessary, the material strained before it is used.

Workmanship should be of a high standard. Painting and finishing should be done by skilled mechanics using proper types and sizes of brushes, roller covers, and spray equipment.

Equipment should be kept clean and in proper condition. The same rollers or brushes used to paint concrete or masonry should not be used on smooth surfaces. Rollers used for a gloss finish and the corresponding primer should have a short nap.

Material should be applied evenly and uniformly under adequate illumination. Surfaces should be completely covered and should be smooth and free from runs, sags, holidays, clogging, and excessive flooding. Completed surfaces should be free of brush marks, bubbles, dust, excessive roller stipple, and other imperfections. Where spraying is required or permitted, the paint should be free of streaking, lapping, and pile-up.

The number of coats recommended by the manufacturer for each particular combination of use, substrate, and paint or transparent finish system should be the minimum number used. When properly applied, that number of coats should produce a fully covered, workmanlike, presentable job. Each coat should be applied in heavy body, without improper thinning. Paint primer coats may be omitted from previously primed surfaces, but every coat recommended should be applied to every surface to be covered. When stain, dirt, or undercoats show through the final coat of paint or finish, defects should be corrected and the surface covered with additional coats until the paint or finish film presents a uniform finish, color, appearance, and coverage.

Paints and transparent finishes should be applied at such rates of coverage that the fully dried film thickness for each coat will not be less than that recommended by the manufacturer. Special attention should be given to ensure that edges, corners, crevices, welds, and exposed fasteners receive a dry film thickness equivalent to that of flat surfaces.

The surfaces behind movable equipment are usually painted the same as similar exposed surfaces. Surfaces behind permanently fixed equipment or furniture are usually given just a prime coat before final installation of the equipment.

Paint coats should be brushed or rolled onto concrete, concrete masonry units, plaster, and gypsum board. Paint coats on doors and frames, and on metal, should be applied by brush.

It is necessary to sand lightly between each succeeding enamel or varnish coat. High-gloss paint should be sanded between coats using very fine grit sandpaper. Dust should be removed after each sanding to produce a smooth, even finish.

Door manufacturer's warranties usually require that doors have the top and bottom edges and cutouts of job-finished wood doors sealed with the same type finish as is applied to the face. Regardless of warranty considerations, the tops, bottoms, and side edges of exterior doors should be finished the same as the exterior faces of the doors. The edge and cutout sealer should be applied after the doors have been fitted and before the faces are finished. Sealer should not be allowed to run onto the faces or edges of the doors.

The tops, bottoms, and edges of finish-carpentry items, including shelves, should be finished after the units are fitted, using the same material and methods as used on the exposed surfaces of the units.

Where a transparent finish other than an oil finish will be applied on the interior of a building over an open-grained wood, such as elm, oak, hickory, or walnut, paste wood-filler should be applied and wiped across the grain as it begins to flatten. A circular motion should be used to secure a smooth, filled, clean surface, with filler remaining in the wood's open grain. Where the wood is to be stained, the filler is often slightly tinted with the stain to avoid emphasizing the grain. The filler may also be slightly stained in order to emphasize the grain where the overall surface will not be stained. After the filler has dried overnight, the surface should be sanded until smooth before the next coat is applied. Steel wool should not be used. Where the next coat is a stain, the filler should be allowed to dry for at least 24 hours before the stain is applied.

Surfaces to be stained and given a transparent finish should be covered entirely with a uniform coat of stain, using either a brush or clean cloth. Excess stain should then be wiped off.

When a classic oil finish is desired, the boiled linseed oil to be used should be thinned with turpentine. The mixture should be between equal parts of oil and turpentine to twice as much oil as turpentine. The actual mixture depends solely on the personal preferences of the applicator. An oil finish may be applied over a stain or directly to bare wood. The oil-turpentine mixture should be brushed liberally onto the wood until no more is absorbed. Then the excess should be wiped off with clean cloths and the surface polished to a sheen. Sufficient pressure should be exerted to melt the oil. This can be achieved with an orbital sander or by hand pressure. After a 48-hour drying period, the entire process should be repeated at least five times.

Weather-stripped doors and frames should be finished before door equipment is installed.

The first coat (primer) recommended by the paint manufacturer may be omitted on metal surfaces which have been shop-primed, providing that the primer is touched-up in the field to repair abrasions and other damage.

Primed and sealed surfaces that show evidence of suction spots or unsealed areas in the first coat, should be reprimed to assure a finish coat with no burn-through or other defects due to insufficient sealing.

At plaster to be painted, adjacent trim and accessories should also be properly prepared and painted.

The first coat of paint or transparent finish material should be applied to surfaces that have been cleaned, pretreated, or otherwise prepared for painting or finishing as soon as practicable after preparation and before subsequent surface deterioration. Sufficient time should be allowed between successive coatings to permit proper drying. A minimum of 24 hours is required between interior coats and 48 hours between exterior coats. Exterior coats of cement-based paints, however, require a minimum of only 24 hours between coats. Surfaces should not be recoated until the previous coat has dried until it feels firm, does not deform or feel sticky under moderate thumb pressure, and the application of another coat of paint or transparent finish material does not cause lifting or a loss of adhesion of the undercoat.

The edges of paint adjoining other materials or colors should be made sharp and clean in straight lines, without overlapping.

Mechanical and electrical equipment and devices that have not been factory-finished should be painted as follows. Pumps, fans, and heating and ventilating units should receive two coats of paint. The interior surfaces of ducts and chases, where the surfaces are visible through registers or grilles, should be painted with a flat, nonspecular black paint. Exposed brass or copper pipe should be painted the same as if it were steel and iron. Name plates, moving parts, and polished surfaces should not be painted. Hangers and supports should also be painted while pipes and equipment are being painted.

Paint and transparent finishes should be cured in the proper humidity and temperature conditions, as recommended by the materials manufacturers. Cement-based paints should be kept damp by spraying with water between coats and during the curing period, which requires several days after application of the final coat. Other paints should be kept dry until cured, unless their manufacturer specifically recommends otherwise.

At the completion of painting and finishing, damaged paint and finishes should be examined, touched-up and restored where damaged, and left in proper condition.

Clean-up and Protection.   At the end of each work day, while painting and finishing are being done, discarded paint materials, rubbish, cans, and rags should be removed from the site.

When painting and finishing have been completed, glass and other paint-spattered surfaces should be cleaned. Spattered paint should be removed using scraping or other methods that will not scratch or otherwise damage the finished surfaces.

Paint and Transparent Finish Failures and What to Do about Them

Why Paint and Transparent Finishes Fail

Some paint and transparent finish failures can be traced to one or several of the following sources: structure failure; structure movement; solid substrate problems; supported substrate problems; and other building element problems (Fig. 5-4). Those sources are discussed in Chapter 2. Many of them, while

Figure 5-4 A plumbing leak caused the paint failure shown in this photo. *(Photo by author.)*

perhaps not the most probable cause of a paint or transparent coating failure, may be more serious and costly to fix than the types of problems discussed in this chapter. Consequently, the possibility that they are responsible for a paint of transparent finish failure should be investigated.

The causes discussed in Chapter 2 should be ruled out or, if found to be at fault, repaired before paint or transparent finish repairs are attempted. It will do no good to apply a new paint or transparent finish when a failed, uncorrected, or unaccounted-for problem of the types discussed in Chapter 2 exists. The new installation will also fail. After the problems discussed in Chapter 2 have been investigated and found to be not present, or repaired if found, the next step is to discover any additional causes for the paint or transparent finish failure and correct them. Possible additional causes include bad materials, improper design, bad workmanship, failure to protect the installation, poor maintenance procedures, and natural aging. The following paragraphs discuss those additional causes. Included with each cause is a numbered list of errors and situations that can cause a paint or transparent finish failure. Refer to ''Evidence of Failure'' later in this chapter for a listing of the types of failure to which the numbered failure causes apply.

Bad Materials.   Improperly manufactured paint or transparent finish materials arriving at a construction site is certainly not unheard of, and that possibility should be considered when paint or transparent finishes fail. But, the number of incidents of bad materials is small compared with the cases of bad design or workmanship. Types of manufacturing defects that might occur include:

1. Too much oil in the paint. This can cause alligatoring and checking.
2. Paint or transparent finish materials with inconsistent colors.
3. Paint or transparent finish materials that are manufactured using materials that are inconsistent in color.
4. Using pigments that are incompatible with the other ingredients in the paint.
5. Paint or transparent finish materials that are inconsistent in composition or density.
6. Paint that contains too much pigment for the binder. Such materials will chalk excessively.
7. Paint or transparent finish formulation or defective or old materials which prevent the paint or transparent finish material from drying. Not enough drier in a paint can cause this condition, as can the use of a poor-quality solvent that vaporizes too slowly. A slow-drying oil will also cause paint to dry too slowly. Poorly formulated paint may never dry. Old shellac may dry slowly and may never dry completely.

Improper Design.   Improper design includes selecting the wrong paint or transparent finish material for the location and conditions, as well as requiring an improper installation. The following design problems can lead to paint or transparent finish failure:

1. Selecting paint or transparent finish materials of inappropriate composition for the location and use intended. Using residential-quality paint in a commercial lobby is an example. Selecting latex paint for use where the humidity will be high is another. Very high humidity can cause the water-soluble components in latex paints to appear as brown spots in the dried paint. Using the wrong primer or top coat over plaster can result in uneven dark paint colors, due to uneven absorption of paint. Using paint that is not alkali-resistant over cement plaster can result in unevenness in light-colored paints or flat spots in glossy paints. Using the improper primer can result in uneven colors in paints over gypsum board, due to uneven absorption of the paint by the board's paper surface and the joint compounds. This condition is more likely to occur when oil- or alkyd-based paints are used.
2. Selecting incompatible paint system products. Using a water-based

top coat over a solvent-based undercoat may cause alligatoring, checking, or peeling, for example. Using a hard finish over a soft undercoat can result in alligatoring or checking. Latex paint applied over old chalking oil paint cannot penetrate the chalk and probably will not adhere. Applying an oil paint over a latex paint can cause separation, because when they age the oil paint becomes harder and less elastic than the latex. Incompatible top coats may also blister.

3. Selecting paints or transparent finish materials that are incompatible with the substrates, including existing paints or transparent finishes. Peeling will often result, for example, if paints that are not alkali-resistant, such as most oil- and alkyd-based paints, are used directly on concrete, unit masonry, or cement plaster (stucco), unless such surfaces are pretreated or primed with alkali-resistant primers.

4. Selecting paints or transparent finish materials of inferior quality. An example is selecting a paint that has excessive chalking characteristics. Selecting a chalking paint for use where the runoff will cross another material, such as brick, is another.

5. Failing to require removal of paint that is more than 1/16 inch thick. Thick undercoats can cause the new coats to alligator or crack.

6. Failing to require proper preparation or application methods.

7. Failing to require the proper number of coats.

Bad Workmanship.   Correct preparation and installation are essential if paint or transparent finish failures are to be prevented. The following workmanship problems can lead to paint or transparent finish failure:

1. Failing to follow the design and the recommendations of the manufacturer and recognized authorities. Applying fewer than the recommended number of coats, for example, can result in failure to cover and telegraphing of underlying faults through the paint.

2. Applying paint or transparent finish materials before the building has been completely closed and before wet work, such as concrete, masonry, and plaster, have dried out sufficiently, and where concrete or plaster is not dry and at the proper level of alkalinity. Applying paint to damp or wet surfaces can cause poor adhesion of the paint and blistering. Painting plaster too soon can cause uneven light colors, or light or flat (in glossy surfaces) spots.

3. Applying paint or transparent finish materials when the humidity where the paint or finish is being applied exceeds the level recommended by the paint or finish manufacturer. Excess humidity and dampness can cause paint to dry slowly and to not adhere or form blisters. Properly ventilating spaces where paint or transparent finish materials are being applied not only helps protect the workers from

an accumulation of potentially toxic fumes, but also helps to control humidity and promote proper drying of paint and transparent finish materials. Brown spots may appear in latex paints when the humidity is too high.

4. Installing paint or transparent finish materials when the room temperature has been less than that recommended by the paint manufacturer for 48 hours before the installation or when the paint material's temperature is too low. The manufacturer's recommended minimum temperature may range from 45 to 65 degrees Fahrenheit, and will vary also with paint type. An associated problem is letting the room temperature fall below the temperature recommended by the manufacturer after the paint has been applied. Low temperature is a major factor in paint failures, because it can affect drying time and adhesion. Low temperatures shorten drying time and make the paint hard to apply. Applying paint containing ice and applying paint to a frozen or frost-covered surface can lead to poor adhesion and paint blistering. Low temperatures can cause paint wrinkling and may cause the paint to not adhere.

5. Applying paint when the temperatures of the air, surface being painted, or paint is too high. High temperature is a major factor in paint failures. It can thin the paint and make it not cover as well as it otherwise would. Paint that is too warm sets too rapidly and forms too thin a film.

   Hot air can also dry the surface too rapidly and trap wet paint beneath it. The result can be bubbles that may rupture, forming pits or puncture wounds that can extend completely through the paint to the substrate.

   Applying paint in direct sunlight or when the air temperature is too high may cause the surface of the paint to dry too quickly, entrapping solvent vapors that will appear as blisters.

   Paint, surfaces, or air that is too hot can also cause the paint surface to wrinkle or the paint to not adhere.

   The manufacturer's recommended high temperatures for paint application may range from 85 degrees Fahrenheit to 120 degrees Fahrenheit, and may vary also with the paint type.

6. Permitting drastic changes in temperature. When drastic temperature changes occur before a paint has completely set, alligatoring or checking may occur when the paint expands or contracts. If the top coat is not elastic enough, it will crack.

7. Permitting a wide temperature differential to exist between the paint and the surface to which it is applied. Large differences may cause the paint to blister and eventually peel.

8. Failing to properly prepare the area where a paint or transparent fin-

ish material will be installed, removing mildew, moss, ivy and other plant growth (Fig. 5-5), efflorescence, laitence, oil, grease, dirt, dust, rust, corrosion, loose mill scale, wax, calcimine, loose existing paint, and other contaminants that will interfere with proper application, damage the paint or transparent finish materials, or telegraph through the paint or transparent finish material.

Paint or transparent finish materials applied over grease or oil may dry slowly or not at all.

Improperly prepared metal, such as nails and flashings, will rust or corrode when they contact moisture and stain the paint. Iron stains are brown. Copper and bronze stains are bluish-green.

Unremoved previous coats of calcimine wash will cause paint to peel, flake, crack, or scale.

Necessary preparation includes removing or spot priming harmful materials embedded in the substrates. For example, brown spots in paint over plaster may be caused by grease or silt embedded in the plaster. Where such materials occur, the area should be scraped and sanded thoroughly so that it is as smooth as possible. The spots and the sanded and scraped area should then be covered with two

Figure 5-5 The ivy shown in this photo must be completely removed before repainting can begin. (*Photo by author.*)

coats of aluminum paint or shellac. When dry, the surface can be painted.

Preparation also includes smoothing rough surfaces and roughening surfaces that are so smooth that paint or transparent finishes will not adhere, and ensuring that surfaces are dry and that no other condition exists that is detrimental to the formation of a durable paint or finish film.

9. Failing to remove sources of moisture and water that will affect the paint or transparent finish and to ensure that the substrate is completely dry. Water and moisture may cause the paint or finish to dry slowly, and will aid mildew, moss, and other plant growth, and cause the paint or finish to blister and eventually peel, flake, crack, or scale.

   Alligatoring in shellac is almost always caused by exposure to damp conditions.

10. Failing to remove dust, dirt, moisture, and other contaminants between coats. This can cause loose, peeling, cracking, flaking, or scaling paint. Moisture left between coats can cause the paint to not dry properly and may cause blisters which eventually lead to interlayer peeling.

11. Failing to neutralize alkali in masonry substrates or use alkali-resistant paints. Either can cause peeling. An alkali residue can cause a paint's gloss to be poor or inconsistent.

12. Failing to remove natural salts from previously painted surfaces. Salts which normally collect on painted surfaces are washed off by rain where exposed. They must be removed from areas which rain does not touch by washing with trisodium phosphate. Failing to remove salts will result in peeling paint.

13. Failing to remove loose, peeling, or otherwise unsound paint or transparent finish materials from an existing surface before applying new paint or transparent finish materials.

14. Applying paint or transparent finish materials over a slick or glossy surface. Failing to remove the gloss from an existing surface or undercoat before applying a succeeding coat of paint or transparent finish to it will result in a failure of the paint or finish to adhere. Applying new paint over existing glossy paint without roughening the surface is an example. Applying paint directly over ceramic tile, structural facing tile, or glazed brick (Fig. 5-6) are other examples.

15. Failing to sand previous coats before the next coat is applied.

16. Failing to prime water-stained surfaces before painting.

17. Failing to remove chalking before repainting.

18. Failing to remove rusted nails and other fasteners, and other rusted items, or to clean the rust off and apply a rust-inhibitive primer.

Figure 5-6 This paint, applied directly over glazed brick, never had a chance of succeeding. (*Photo by author.*)

19. Failing to countersink nails and screws, fill holes, and spot prime.
20. Failing to remove discolorations caused by color extractives in wood substrates, and to provide a stain-blocking primer.
21. Failing to clean and apply a knot primer to knots that are already bleeding, or may bleed in the future.
22. Removing existing paint using inappropriate methods. Open flames may char the wood. Rotary sanders, sandblasting, and waterblasting may remove too much material or otherwise damage some surfaces, particularly wood.
23. Failing to sand shoulders at the edges of sound paint or otherwise feather the edges of existing paint where new paint is applied.
24. Roughening gypsum board joints or plaster surfaces by sanding, or painting over surfaces so roughened. This can cause uneven absorption of paint between the paper facing and the joint compound or between the sanded and unsanded plaster, resulting in uneven colors.
25. Failing to properly prepare wood substrates, including failing to remove residue from knots, pitch streaks, cracks, open joints, and sappy spots; to apply a coat of white shellac to pitch and resinous sapwood before the prime coat is applied; to scrape and clean small,

dry, seasoned knots and apply a thin coat of white shellac or other knot sealer to them before the prime coat is applied; and to sand surfaces before painting or finishing.

26. Failing to properly prepare metal substrates.
27. Failing to fill holes and imperfections in the substrate.
28. Using paint or transparent finish materials that are of the wrong type for the installation.
29. Applying paint or transparent finish materials that are incompatible with previous coats, existing paint or finish materials, or substrates.
30. Applying damaged paint or transparent finish materials, regardless of whether the damage was inherent in the manufactured materials or occurred during shipment, storage, or installation.
31. Applying paint or transparent finish on materials that are stained.
32. Using the wrong primer. This can cause many problems. Over gypsum board it can cause uneven absorption of paint and uneven colors in the top coat.
33. Using the wrong, defective, or poor-quality thinner. Such solvents may evaporate too quickly, resulting in paint wrinkles, alligatoring, blistering, peeling, flaking, cracking, or checking. Or they may dry slowly, resulting in a paint or finish that dries too slowly, or never dries completely.
34. Using old shellac. Old shellac may not dry.
35. Using too little thinner. This can result in peeling, flaking, cracking, or scaling.
36. Using too much thinner. Thin paint over gypsum board surfaces may not be equally absorbed by the paper covering and the joint compounds, resulting in uneven paint colors. This is more of a problem with alkyd- or oil-based paints than with water-based materials.

    Adding too much thinning material in order to save paint may result in premature aging, fading, or the inability of the paint to withstand normal cleaning.
37. Failing to properly agitate and mix paint. Improper mixing before and during application can cause color separation. Insufficient mixing resulting in the pigments not being properly blended may cause alligatoring, checking, peeling, flaking, cracking, or scaling. Improperly mixed paint may not dry.
38. Mixing incompatible paints.
39. Adding lampblack or another shading material to professional paint materials to make the paint hide better, with less paint, can result in the paint failing to properly adhere, changing in color, or prematurely aging.
40. Adding too much oil to paint can cause alligatoring or checking, and may cause the paint to dry improperly.

41. Using pigments that are incompatible with other ingredients can result in alligatoring, checking, peeling, flaking, cracking, or scaling.
42. Painting too soon over green (uncured) plaster. This can result in peeling, flaking, cracking, or scaling.
43. Applying paint to gypsum board before joint compounds are dry.
44. Failing to properly apply paint or transparent finish materials, including not using enough material, improperly applying the material, and not applying the correct number of coats.
45. Failing to completely cover the surface.
46. Applying paint coats that are too thin.
47. Applying paint in coats that are too thick. When the surface film forms before the paint beneath has dried, the film is smooth across the surface. When the underlying paint dries, it shrinks, wrinkling the surface film.

    Paint coats that are too thick may also result in alligatoring and checking.

    Too thick a paint coat, especially an undercoat, can cause peeling, flaking, cracking, or scaling.

    Paint or transparent finish materials that are applied in too-thick coats may run or sag, and may not properly dry.
48. Applying a coat of paint to a still-wet undercoat. This can cause the paint to dry slowly, or to wrinkle. It may also cause alligatoring, checking, blistering, peeling, flaking, cracking, or scaling.
49. Applying a hard finish coat over a relatively soft undercoat. This can cause the paint to wrinkle, and may result in alligatoring, checking, or peeling of the top coat.
50. Failing to brush out paint. This can cause wrinkling, runs, or sags.
51. Entrapping air in the paint. This may result from too much air pressure in spray paint application, which can cause air bubbles to become trapped beneath the paint's surface.
52. Permitting water in the air line used in spray applications.
53. Improperly adjusting the spray gun pattern, applying sprayed paint using too high an air pressure. These can cause wrinkled paint.
54. Spraying paint when there is grease or oil in the spray line.
55. Improperly applying paint using spray equipment.

Failing to Protect the Installation.   Paint or transparent finishes must be protected before, during, and after installation. Errors include the following:

1. Failing to protect paint or transparent finish materials from staining or marring by other construction materials.
2. Permitting contaminants, including airborne dust from broom cleaning while the paint or finish is still wet, to fall onto wet, newly painted or finished surfaces.

3. Permitting rain to fall on wet paint. Rain can pit the surface of any paint and may remove the gloss from oil- or alkyd-based paint, or wash the paint from the wall.
4. Permitting abuse before or during installation or after the paint or transparent finish material has been applied.
5. Permitting the temperature in a space where paint or transparent finishes have been installed to fall below 55 degrees Fahrenheit or rise above 95 degrees Fahrenheit. High or low temperatures cause expansion and contraction in the paint or finish and substrates. Extremes may cause the paint or finish to crack, especially if they are old or naturally hard, or for some other reason are not elastic enough to withstand the stresses involved.
6. Permitting deterioration of the substrates, flashings, or other construction to occur, so that the substrates become wet or corrode.

Poor Maintenance Procedures.   Not maintaining paint or transparent finishes properly can result in failures. Poor maintenance procedures include:

1. Failing to clean paint or transparent finishes regularly. Unremoved grease, dirt, and other contaminants can result in permanent stains.
2. Improper cleaning of paint or transparent finishes. Paint or transparent finishes may be damaged severely by sandpaper, steel wool, and abrasive or caustic cleaners.
3. Failing to keep vines and other vegetation from growing on or close to building surfaces. Vines can penetrate the surface (Fig. 5-7), which damages the surface and permits water to enter. Water in the substrates can cause much damage to the substrates and their paint coatings. Paint may blister or peel, for example. Mildew and moss may grow on the wet surfaces. Water may force efflorescence from the substrate causing the paint to peel or flake off. Water in a wood substrate can set up a chemical reaction between the water and the natural extractives in the wood, staining the paint.
   Vegetation close to the walls can shade them and make an environment conducive to mildew and moss growth.

Natural Aging.   All paint and transparent finishes age and lose their properties over time. The manufacturer is the best source of data about the expected life of a product. It should be kept in mind, however, that the manufacturer's projections may assume the best of conditions. Paint and transparent finishes may fail for the following causes:

1. Failing to repaint or refinish at proper intervals. Paints and transparent finish materials which are exposed to the sun's ultraviolet rays

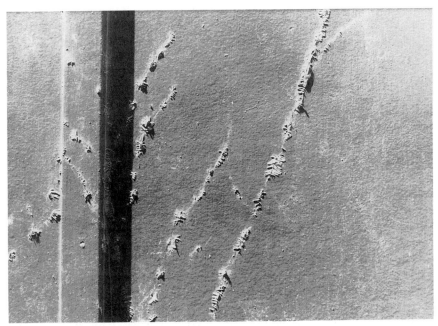

Figure 5-7 The contamination on the exterior gypsum panels shown in this photo are ivy roots. The ivy is dead and gone, but the roots remain locked to the surface and difficult to remove. (*Photo by author.*)

will fail faster than those used in other locations. All paints will fail eventually if not maintained and repainted at the proper intervals (Fig. 5-8).

Exterior wood with a transparent finish is particularly susceptible to problems because the finish, being clear, does not prevent the sun's rays from reaching the wood. In addition, the ultraviolet rays may also attack the finish itself. The result can be delamination of the finish from the surface, flaking, and peeling. Exterior transparent finishes containing alkyd resins may fail in as little as six months. Urethane resins may last eighteen months. Even the best materials, phenolic resins with ultraviolet absorbers, may last only as long as two years.

2. Bleeding of natural finished wood. Unprotected wood exposed to weather will turn dark over time as water-soluble impurities bleed out. The appearance of bleeding on transparent-finished exterior wood surfaces indicates that the finish has failed and that water has

Figure 5-8 No paint lasts forever. (*Photo by author.*)

reached the substrate. The worst woods for bleeding are redwood and cedar.
3. Mildew growth, due to a failed finish permitting water to reach the substrate.
4. Loss of elasticity. Old paint that has lost its elasticity becomes hard and is no longer able to respond to the expansion and contraction of the material to which it is applied. The result is crazing and eventually alligatoring, flaking, or peeling.
5. Natural erosion. Weathering paint will slowly wear away until it eventually does not properly protect the surface painted (Fig. 5-9).

Evidence of Failure

In the following paragraphs, paint or transparent finish failures are divided into failure types, such as "Cracked Paint." Following each failure type are one or more failure sources, such as "Improper Design." After each failure source, one or more numbers is listed. The numbers represent possible errors associated with that failure source, which might cause that failure type to occur.

Figure 5-9 The owner waited too long to have this fence repainted. (*Photo by author.*)

A description and discussion of the numbered failure causes for failure types "Steel and Concrete Structure Failure," "Wood Structure Failure," "Structure Movement," "Solid Substrate Problems," "Supported Substrate Problems," and "Other Building Element Problems" appears under those headings in Chapter 2. Each of those headings is listed in the Contents.

The following example applies to the six failure types listed in the preceding paragraph. Clarification and explanation of the numbered cause (2) in the example

■ Solid Substrate Problems: 2 (see Chapter 2).

appears under the subheading "Solid Substrate Problems" in the part of Chapter 2 called "Substrates," where Cause 2 reads

2. The solid substrate material cracks or breaks up, joints crack, or surfaces spall due to bad materials, incorrect material selection for the location and application, or bad workmanship.

A description and discussion of the types of problems and numbered failure causes that follow failure types "Bad Materials," "Improper Design," "Bad Workmanship," "Failing to Protect the Installation," "Poor Maintenance Procedures," and "Natural Aging" appears under those headings earlier in this chapter under the main heading "Paint and Transparent Finish Failures and What to Do about Them," subheading "Why Paint and Transparent Finishes Fail," which is listed in the Contents. For example, clarification and explanation of the type of problem (Improper Design) and numbered cause (2) in the example

■ Improper Design: 7.

appears earlier in this chapter under the main heading "Paint and Transparent Finish Failures and What to Do about Them" subheading "Why Paint and Transparent Finishes Fail," sub-subheading "Improper Design," Cause 7, which reads:

7. Failing to require the proper number of coats.

Paint failure types include the following.

Stained or Discolored Paint or Transparent Finish.   Blemishes include stains, brown spots, and other discolorations; differences in color or sheen within a surface; mildew, moss, and other plant growth; and bleeding knots. One or more of the following may be at fault:

■ Solid Substrate Problems: 1 (see Chapter 2).
■ Supported Substrate Problems: 5, 7, 13, 17 (see Chapter 2).
■ Other Building Element Problems: 1, 2 (see Chapter 2).
■ Bad Materials: 2, 3, 4, 5.
■ Improper Design: 1, 3, 6.
■ Bad Workmanship: 1, 2, 3, 8, 9, 10, 11, 16, 18, 19, 20, 21, 24, 25, 28, 30, 31, 32, 36, 37, 38, 39, 41, 42, 43.
■ Failing to Protect the Installation: 1, 2, 4, 6.
■ Poor Maintenance Procedures: 1, 3.
■ Natural Aging: 2, 3.

Wrinkling.   Wrinkles appear in paint when the surface dries before the underlying paint has dried. It can result from one or more of the following causes:

- Bad Materials: 7.
- Improper Design: 2, 6.
- Bad Workmanship: 1, 4, 5, 6, 33, 47, 48, 49, 50, 53, 55.

Alligatoring and Checking.   Cracks that resemble an alligator's hide are called alligatoring (Fig. 5-10). Checking is similar, except that the cracks are smaller and not as noticeable. In their early stages, these defects are

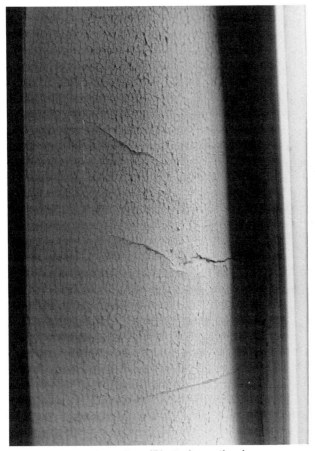

Figure 5-10   Alligatoring. (*Photo by author.*)

also sometimes called crazing, crowfooting, or hairlining. They are caused by the top coat being unable to properly adhere to underlying coats. In their early stages they affect only the surface coat, but later the cracks may extend through to the substrate. They result from one or more of the following causes:

- Wood Structure Failure: 3 (see Chapter 2).
- Structure Movement: 1, 2, 3, 4, 5, 6, 7, 8 (see Chapter 2).
- Solid Substrate Problems: 2, 3, 4 (see Chapter 2).
- Supported Substrate Problems: 1, 2, 3, 9, 10 (see Chapter 2).
- Bad Materials: 1, 4, 5.
- Improper Design: 1, 2, 4, 5, 6.
- Bad Workmanship: 1, 6, 9, 10, 29, 30, 33, 35, 37, 38, 40, 41, 44, 47, 48, 49.
- Failing to Protect the Installation: 5.
- Natural Aging: 1, 4.

Peeling, Flaking, Cracking, and Scaling.   Peeling, flaking, cracking, and scaling may be advanced stages of alligatoring (Fig. 5-11), checking, or blistering, or may occur independently. Paint may peel completely away from the substrate (Fig. 5-12) or outer layers may peel away from undercoats. When cracks extend completely through a paint's film down to the substrate, water works its way behind the paint film and causes it to flake (small pieces) or scale off (Fig. 5-13). Scaling is an advanced form of flaking. These conditions may be due to one or more of the following causes:

- Wood Structure Failure: 3, 5 (see Chapter 2).
- Structure Movement: 1, 2, 3, 4, 5, 6, 7, 8 (see Chapter 2).
- Solid Substrate Problems: 1, 2, 3, 4 (see Chapter 2).
- Supported Substrate Problems: 1, 2, 3, 4, 6, 9, 10, 11, 12, 15, 16 (see Chapter 2).
- Other Building Element Problems: 1, 2 (see Chapter 2).
- Bad Materials: 1, 4, 5.
- Improper Design: 1, 2, 3, 4, 5, 6.
- Bad Workmanship: 1, 2, 3, 4, 5, 6, 7, 8, 9, 10, 11, 12, 13, 14, 15, 17, 26, 28, 29, 30, 32, 33, 35, 37, 38, 39, 40, 41, 42, 43, 44, 47, 48, 49, 54.
- Failing to Protect the Installation: 5, 6.
- Poor Maintenance Procedures: 3.
- Natural Aging: 1, 4.

Figure 5-11 The alligatored paint on this overhead door, left uncorrected, eventually flaked off, exposing the wood substrate, which got wetter, causing even more flaking. (*Photo by author.*)

Blistering.   Blisters may appear as pimples, bubbles, pinholes, or pits in the paint. Blisters appear when vapor, either solvent or water, is trapped beneath a paint's impermeable surface. There are several reasons why paint may blister, including the following causes:

- Wood Structure Failure: 5 (see Chapter 2).
- Solid Substrate Problems: 1 (see Chapter 2).
- Supported Substrate Problems: 4, 15 (see Chapter 2).
- Other Building Element Problems: 1, 2 (see Chapter 2).
- Bad Materials: 7.
- Improper Design: 2, 3, 6.
- Bad Workmanship: 1, 2, 3, 4, 5, 7, 9, 10, 29, 33, 48, 51, 52, 53, 55.
- Failing to Protect the Installation: 6.
- Poor Maintenance Procedures: 3.

Figure 5-12 This photo shows as bad a case of peel-ing paint as anybody has ever seen. (*Photo by author.*)

Excessive Chalking.   When the resin in a paint film disintegrates, chalking, which is a powder on the surface, is the result. Some chalking is desirable and is the natural weathering method of chalking paints. When rain washes the chalking off, dirt and other soiling goes with it. Excessive chalking, however, can weaken the paint coating as it becomes thinner, eventually resulting in a coating too thin to protect the underlying surface. Excessive chalking may result from one or more of the following causes:

- Bad Materials: 6.
- Improper Design: 1, 4.

Figure 5-13 The cracks in the paint on the window stool in this photo were caused by condensation running down the face of the window and standing on the stool. The cracked paint has begun to flake. (*Photo by author.*)

Paint Wash-off. When rain strikes a paint surface that has not dried or has not dried properly, the outer layer can be washed away. This condition can result from one or more of the following causes:

- Other Building Element Problems: 1 (see Chapter 2).
- Bad Materials: 1, 7.
- Bad Workmanship: 3.
- Failing to Protect the Installation: 3.
- Poor Maintenance Procedures: 2.

Paint Erosion. Paint wears away (Fig. 5-14) due to the following causes:

- Other Building Element Problems: 1 (see Chapter 2).
- Failing to Protect the Installation: 3.
- Poor Maintenance Procedures: 2.
- Natural Aging: 5.

Figure 5-14 The overhead door in this photo needs care badly if further damage is to be prevented. Repainting is in order soon. (*Photo by author.*)

Failure to Dry as Rapidly as Expected. Paint may remain tacky for extended periods and dry too slowly due to one or more of the following causes:

- Wood Structure Failure: 5 (see Chapter 2).
- Solid Substrate Problems: 1 (see Chapter 2).
- Other Building Element Problems: 1, 2 (see Chapter 2).
- Bad Materials: 1, 7.
- Improper Design: 4, 6.
- Bad Workmanship: 1, 2, 3, 8, 9, 10, 25, 30, 33, 34, 37, 38, 40, 44, 47, 48, 54.
- Failing to Protect the Installation: 3, 5.

Failure to Ever Dry. Good paint or natural finish materials will eventually dry regardless of the application methods used or the climatic conditions.

When a paint or natural finish never dries, it is for one of the following reasons:

- Bad Materials: 7.
- Bad Workmanship: 1, 34, 37.

Poor Gloss.   The failure of a paint to achieve or maintain the expected level of gloss may be due to one of the following causes:

- Solid Substrate Problems: 1 (see Chapter 2).
- Supported Substrate Problems: 17 (see Chapter 2).
- Other Building Element Problems: 1, 2 (see Chapter 2).
- Bad Materials: 5.
- Improper Design: 1, 3, 4, 6.
- Bad Workmanship: 1, 2, 3, 4, 6, 8, 10, 11, 24, 28, 30, 33, 36, 37, 38, 39, 44, 45, 46, 48.
- Failing to Protect the Installation: 1, 2, 3, 4.
- Poor Maintenance Procedures: 1, 2.
- Natural Aging: 1, 5.

Photographing or Shadowing.   Material or soiling underneath the paint may photograph through the paint, or not be properly hidden. Underlying structure may show as dark lines or areas called shadowing, due to one of the following causes:

- Wood Structure Failure: 5 (see Chapter 2).
- Supported Substrate Problems: 17 (see Chapter 2).
- Other Building Element Problems: 5 (see Chapter 2).
- Improper Design: 1, 3, 4, 6, 7.
- Bad Workmanship: 1, 5, 8, 16, 18, 19, 20, 21, 22, 23, 24, 25, 27, 28, 32, 33, 36, 37, 43, 44, 45, 46, 55.
- Failing to Protect the Installation: 6.

Runs and Sags.   Paint may not present a smooth surface due to the following causes:

- Bad Workmanship: 1, 4, 5, 6, 7, 36, 44, 47, 50, 55.

Cleaning and Repairing Paint and Transparent Finishes

General Requirements.   The extent of cleaning and repairing paint or transparent finish materials depends on the type and extent of the soiling or damage. This section discusses cleaning and repairing needed when the damage is minor. "Installing Paint and Transparent Finishes over Existing Materials" later in this chapter discusses major extensions of existing paint and transparent finish materials, and the installation of new paint and transparent finishes over existing materials, including existing paint and finish removal and preparation of the substrates to receive the new paint or finish.

The following paragraphs contain some suggestions for cleaning and repairing paint and transparent finish materials. Because the suggestions are meant to apply to many situations, they might not apply to a specific case. In addition, there are many possible cases that are not specifically covered here. When a condition arises in the field that is not addressed here, advice should be sought from the additional data sources mentioned in this book. Often, consultation with the manufacturer of the materials being cleaned or repaired will help. Sometimes, it is necessary to obtain professional help (see Chapter 1). Under no circumstances should the specific recommendations in this chapter be followed without careful investigation and the application of professional expertise and judgment.

The discussion earlier in this chapter about new paint or transparent finish materials and installations also applies to new paint or transparent finishes used in repairing existing surfaces. There are, however, a few additional considerations to address when the new materials are applied in an existing space. The following paragraphs address these concerns. Some, but not all, of the requirements discussed earlier are repeated here for the sake of clarity. It is suggested that the reader refer to earlier paragraphs for additional data.

The requirements listed later in this chapter under "Applying Paint or Transparent Finishes over Existing Materials" apply. Existing materials should be removed, substrates should be prepared, and new materials applied in accordance with the paint or transparent finish material manufacturer's recommendations. Only experienced workers should be used to prepare for and apply paint or transparent finish materials.

Before an attempt is made to clean or repair existing paint or transparent finishes, the existing materials manufacturers' brochures for products, installation details, and recommendations for cleaning and repairing should be available and referenced. It is necessary to be sure that the manufacturers' recommended precautions against the use of materials and methods which may be detrimental to the paint or finish materials are followed.

It is difficult, and sometimes impossible, to actually repair paint or transparent finishes in the true sense of the word. Small patches and repairs are almost always obvious and are seldom acceptable for that reason. Therefore, when this chapter uses the word repair with regard to paint or transparent finishes, it almost always means to remove or properly prepare the existing material and cover it with a new coat of material.

Areas where repairs will be made should be inspected carefully to verify that existing materials that should be removed have been removed and that the substrates and structure are as expected and are not damaged. Sometimes substrate or structure materials, systems, or conditions are encountered which differ considerably from those expected. Sometimes unexpected damage is discovered. Both damage that was previously known and damage found later should be repaired before paint or transparent finishes are repaired. Cleaning and repairing should be done, substrates should be prepared and the repairs made strictly in accordance with the material manufacturer's recommendations. Repairs should be made only by experienced workers.

Materials to be used for making repairs should not be permitted to freeze. Spaces should be kept above 55 degrees Fahrenheit before and after the repairs are made.

Adequate ventilation should be provided while work associated with paint or transparent finish materials preparation and application are in progress. This may be somewhat more difficult when operating in an existing building, especially where occupants are involved. The paint or transparent finish materials manufacturer's recommended safety precautions should be followed.

The moisture content of the substrates should be within the range recommended by the paint or transparent finish materials manufacturer, especially if excess moisture contributed to the failure that is being repaired.

Structural Framing, Substrates, and Other Building Elements.   While improperly designed or installed, or damaged steel, concrete, or wood framing, solid substrates, supported substrates, (Fig. 5-15), or other building elements, which are discussed in Chapter 2, may be responsible for paint or transparent finish failures, repairing them is beyond the scope of this book. They should be investigated as a possible source of the failure and repaired as necessary. The discussions in this chapter assume that when those items have been the cause of a failure, they have been repaired and present satisfactory support systems and conditions for the materials being repaired.

Materials and Manufacturers.   Paint, transparent finish, and associated materials used in making repairs should be as listed in the part of this chapter called "Paint and Transparent Finish Materials," and the discussion earlier about manufacturers also applies here.

In addition to new materials being compatible with each other within each paint or transparent finish system, as is necessary in new buildings, new materials for use in existing buildings must be selected for their compatibility with existing materials that will remain in place. Tests should be conducted in the building to ensure compatibility. Incompatible materials should not be used.

Figure 5-15   This failure is due to the plywood substrate delaminating. It is not a paint failure. (*Photo by author.*)

Materials should be delivered to the project site in the manufacturer's original packaging. Materials should be packed, stored, and handled carefully to prevent damage.

Paint and Transparent Finish Systems.   The discussion of paint and transparent finish systems earlier in this chapter applies to systems to be used in making repairs, with several differences. The materials should match those in the original application. Some coats may be eliminated, however. For example, the primer may not be necessary when the damage affects only the finish coats and the damage can be eliminated without removing or damaging the existing undercoats.

Surfaces Usually Repaired.   Any painted or transparent finished surface may need repair. Where major damage is present, the discussion under "Applying Paint or Transparent Finishes over Existing Materials" applies. Here we will discuss only minor repairs to previously painted or finished surfaces. Even minor repairs to paint or transparent finishes means that a portion of the surface must be removed and new materials applied.

Preparation for Repairing Paint and Transparent Finishes. Surfaces to be repaired should be prepared as discussed for major surfaces to be repainted or recoated under "Applying Paint or Transparent Finishes over Existing Materials" and the requirements suggested there generally apply.

Before repairs are attempted, the technique and materials to be used should be applied in a small inconspicuous area to test the compatibility of the materials and matching of the existing color and finish.

Inspecting and Testing. Before a decision can be made about whether to repair a damaged surface or repaint or refinish the surface, it is necessary to conduct an inspection of the surface. Some testing may also be necessary. Some painting contractors are able to conduct such examinations and tests, but many are not. Most national paint manufacturers employ personnel who are trained and equipped to make inspections and tests and to recommend methods of dealing with paint failures.

Tests that may be required include a compatibility test, paint thickness tests, and moisture content tests. The results often point directly to the cause of a failure. Test results can also help prevent further failures caused by misdiagnosis or using the wrong methods to make repairs or repaint failed surfaces. Examination may show, for example, whether just a surface coat is peeling or whether the entire thickness of the paint is effected. Microscopic examination may be necessary to determine the number of coats already applied. Chemical testing may be needed to determine the type of paint (oil-based or water-based, for example) used in the existing coats.

Even before tests are made, it may be possible to determine the cause of some paint failures. Where peeling or chipping has occurred, a piece of the paint from the failed surface or the substrate from which paint has peeled can be examined. Slick surfaces may mean that the substrate was too smooth and must be sanded or otherwise roughened before new paint will adhere properly. The back of the paint chip may also be slick if the substrate was coated with oil or grease when the paint was applied. A chalky residue on masonry or plaster might be efflorescence. Chalky material on the back of paint that was on metal may be rust. White rust will form even on zinc-coated metal. Paint that was over a galvanized surface that is coated with a waxy film may indicate that an oil-based primer was used.

When blistering has occurred, diagnosis can be made by slitting a blister. If the bare substrate is visible, the blister was probably caused by entrapped moisture vapor. If there is paint beneath the blister, the problem was caused by solvent vapor that could not escape into the atmosphere because the paint's surface skin formed too rapidly.

Repairing, Cleaning, Clean-up, and Protection.   The requirements discussed later in this chapter under "Applying Paint or Transparent Finishes over Existing Materials" generally apply to repairs.

Existing surfaces that are to be cleaned and not repainted or refinished should have soiling, stains, dirt, grease, oil, and other unsightly contaminants removed completely. Detergents and cleaning compounds should be compatible with existing surfaces and should not damage existing finishes. Materials and methods to be used should be those recommended by the paint or finish manufacturer. Surfaces damaged due to the use of incorrect products or due to improper use of cleaning agents should be refinished. Dirt, soot, pollution, cobwebs, insect cocoons, and the like can be removed from exterior surfaces using a water spray followed by scrubbing with a mild household detergent and water solution. The cleaned surfaces should be rinsed thoroughly with clean water and permitted to dry.

Existing paint to be repainted should be scrubbed using household cleaner and water, rinsed thoroughly with clean water, and allowed to dry before painting. Mildew, stains, and other blemishes should be treated as discussed under "Applying Paint or Transparent Finishes over Existing Materials."

The extent of painting and finishing of repairs may be just the repair itself, but more normally would extend to natural breaks, such as corners or projections.

Existing transparent finishes are seldom only partly refinished, however. Usually such surfaces are completely refinished. Where a surface cannot be refinished without affecting edges and projections, the edges and projections are usually also refinished.

Following are some recommendations for dealing with common paint and transparent finishes problems. They should, of course, be adapted to suit actual conditions.

Brown Spots on Painted Surfaces.   The method of repair depends on the cause of the spots. Brown spots on plaster are sometimes caused by grease or silt embedded in the plaster. The area should be scraped and sanded thoroughly so that it is as smooth as possible. The spots and the sanded and scraped area should then be covered with two coats of aluminum paint or shellac. When dry, the surface should be painted to match the surrounding area.

When the brown spots are caused by using a latex paint in an area where high humidity is present, the spots should be cleaned away with a damp sponge. Abrasive cleaners should not be necessary. Repainting is not usually needed.

Mildew.   It is first necessary to establish that the stain is actually mildew (Fig. 5-16). A simple test is to place a drop of household bleach on the stain. Mildew will turn white. Dirt will not. Mildew may be removed using the formulation mentioned earlier. Prevention of a reoccurrence may be more difficult, however. Mildew-resistant paints may help. The most effective prevention technique is to remove the cause of the shade and dampness that permitted the growth in the first place. Vegetation may be trimmed to permit sunlight to strike the surface, for example. Sources of water should

Figure 5-16 The stains on the neglected door and frame shown in this photo are mildew, but such is not apparent at first glance. They could just as easily have been dirt. (*Photo by author.*)

be controlled or removed. Adding a rain gutter and downspout system will sometimes help.

Paint Chalking.   Slight chalking can be removed by lightly brushing the surface. Sometimes, light chalking on paint over wood will succumb to washing with a household detergent and water and soft brushes. Heavy chalking can be removed by scrubbing with a solution of trisodium phosphate and water, using a stiff brush. Care should be taken to not damage the surface.

Soft Gummy Paint.   Paint will sometimes become soft and gummy when the substrate is wet. The source of the wetting must be removed or controlled before paint repairs can be accomplished. When the water source cannot be completely eliminated, coating the surface with a cement-based water-proofing sealer before repainting may prevent the condition from recurring.

Wrinkles in Paint.   When the condition is minor, it may be possible to scrape and sand down the wrinkled coat and repaint. More severe conditions may require complete removal of all paint down to the bare substrate before repainting.

Crazing in Painted Surfaces.   An early stage of alligatoring, crazing can be sanded and repainted without complete removal of the paint. The tiny cracks will be apparent on examination, but will be filled with new paint and will, therefore, protect the substrate. If that appearance is not acceptable, complete removal of the crazed paint is necessary.

Alligatoring and Checking in Paint.   Since checking and alligatoring often affect only the surface coats, some sources say that it is sometimes possible to sand or scrape away only the affected coats before repainting. The Brick Institute of America recommends, however, that alligatored paint over brick be completely removed before repainting. Other sources also maintain that complete removal is the only solution to paint alligatoring and checking over wood. When the cracks extend completely through the paint, complete removal is always necessary.

Cracking and Scaling of Paint.   Since cracking and scaling affect the entire thickness of the paint coating, it is necessary to completely remove paint with such damage before repainting. Since the defect was possibly caused by the paint being too brittle for the surface, the new paint should be more flexible in nature.

Blistering of Paint.   Blistered paint caused by moisture in the substrates must be removed completely down to the bare substrates, and the substrates allowed to dry before repainting.

Blistered paint caused by solvent vapors may be removed only down to sound paint.

Peeling of Paint.   When paint has peeled away from the substrates (Fig. 5-17), complete removal is necessary. Since most such peeling is caused by moisture in the substrates, the conditions that caused the peeling must be corrected before the surface can be successfully repainted.

When the surface coat has peeled away from the undercoats, it may be possible to remove the paint only down to sound paint. When the peeling was caused by the use of incompatible paints, the incompatible top coat or coats should be removed completely.

Cracking of Natural Finish.   The cure for cracks in natural finish surfaces depends on the materials used, the extent of the cracking, and the age of the finish. Regardless of the materials used, the first step is to completely remove wax and foreign materials from the finish. Mineral spirits will often

Figure 5-17   The peeling paint in this photo was caused by a plumbing leak in the wall that wet the plaster substrate. (*Photo by author.*)

clean the surface satisfactorily without harming the finish, but any cleaning agent should be first tested in an inconspicuous location to ensure that it will do no harm.

The next step is to test the finish to determine which finish material was used by applying small patches of denatured alcohol and lacquer thinner in inconspicuous locations. The denatured alcohol will soften shellac, but not varnish or lacquer. The lacquer thinner will soften lacquer but not the others. Neither will soften varnish.

Damaged varnish cannot be repaired and must be completely removed and the surface refinished.

Shellac and lacquer can be melted and smoothed out in place. To do so, the appropriate solvent is spread onto the finish, softening the finish material, which can then be brushed out to a smooth finish. It may be best to apply an additional coat of the finish after the refinished material has dried.

When the damage is extensive, or the material has been restored several times, it will probably be necessary to completely remove the existing finish and apply a completely new finish.

Small Worn Spots in Transparent Finishes.   It is often possible to refinish small areas in a transparent finish in such a way that the patch is not objectionably apparent. First, any covering wax or other material must be removed. Rough areas in a wood substrate should be given a coat of penetrating sealer. The new finish should be applied carefully so that lap marks do not show.

Failure of Paint or Natural Finish to Dry.   When a paint or natural finish material does not dry as quickly as expected, the proper approach is to wait. Good materials will eventually dry.

When the paint or natural finish does not eventually dry, the only solution to the problem is to completely remove the paint or natural finish down to the substrate and repaint or refinish using good materials.

Runs and Sags.   Proper application is the best cure. Runs and sags which have been permitted to dry can be removed using sandpaper. Another coat of paint or finish material may be necessary then to achieve the desired finish.

Nailhead Stains.   To remove nailhead stains from painted surfaces or transparent finishes, it is necessary to wire brush the paint or finish from the nails, countersink the nails, prime and fill the holes with wood filler or putty, and repaint or refinish.

Chemical Stains on Painted Wood. Stains caused by chemical reactions between natural extractives in the wood and water can often be cleaned using equal parts of denatured alcohol and water. The cleaned area should then be painted with two coats of a primer that has been specifically formulated for the purpose, after which it can be repainted to match the adjoining surfaces.

Damaged Textured Paint. Since most materials identified as textured paint are really special coatings or textured finishes made from products similar to those used in gypsum wallboard joint compounds, repair of them is beyond the scope of this book.

Damaged sand-finished paint may be repairable, but most likely it will be necessary to first remove at least some of the existing paint. Repairs can sometimes be made using a compatible paint to which a compatible aggregate, perhaps sand, has been added. Matching such surfaces is a difficult job best left to experienced professionals. Even they should practice producing the damaged finish before attempting to make repairs.

Excessive Paint Build-up. Existing paint with a total film thickness of more than 1/16 inch should be removed completely.

Naturally Aged Paint and Transparent Finishes. When paint and natural finishes have reached their expected lifespans, repainting or recoating is necessary. When they are permitted to begin to fail, repainting or refinishing is much more difficult. In some cases, particularly on exterior transparent finished surfaces, complete removal of the existing finish may be necessary.

Applying Paint and Transparent Finishes over Existing Surfaces

Requirements Common to All Paint and Transparent Finishes

The following requirements contain some suggestions for installing new paint or transparent finishes in existing spaces. Because the suggestions are meant to apply to many situations, they may not apply to a specific case. In addition, there are many possible cases that are not specifically covered here. When a condition arises in the field that is not addressed here, advice should be sought from the additional data sources mentioned in this book. Often, consultation with the manufacturer of the materials being installed will help. Sometimes it may be necessary to obtain professional help (see Chapter 1). Under no circumstances should these recommendations be followed without careful investigation and the application of professional expertise and judgment.

The discussion earlier in this chapter about new paint or transparent finish materials and installations also apply to new paint or transparent finishes installed in existing spaces. There are, however, a few additional considerations to address when the new materials are applied in an existing space. The following paragraphs address these concerns. Some, but not all, of the requirements discussed earlier are repeated here for the sake of clarity. It is suggested that the reader refer to earlier paragraphs for additional data.

Substrates should be prepared and new materials applied in accordance with the paint or transparent finish material manufacturer's recommendations. Only experienced workers should be used to prepare for and apply paint or transparent finish materials.

Materials manufacturer's brochures, including installation, cleaning, and maintenance instructions, should be collected for each type of paint or transparent finish material and kept available at the site.

Adequate ventilation should be provided while work associated with paint or transparent finish materials preparation and application are in progress. This may be somewhat more difficult when operating in an existing building, especially where occupants are involved. The paint or transparent finish materials manufacturer's recommended safety precautions should be followed.

Areas where repairs will be made should be inspected carefully to verify that existing materials that should be removed have been removed and that the substrates and structure are as expected and are not damaged. Sometimes substrate or structure materials, systems, or conditions are encountered which differ considerably from those expected. Sometimes unexpected damage is discovered. Both damage that was previously known and damage found later should be repaired before new paint or transparent finish materials are applied.

Structural Framing, Substrates, and Other Building Elements. While improperly designed or installed, or damaged steel, concrete, or wood framing, solid substrates, supported substrates, or other building elements, which are discussed in Chapter 2, may occur, repairing them is beyond the scope of this book. The discussions in this chapter assume that those items have been repaired, if necessary, and present satisfactory support systems and conditions for the new paint and transparent finishes being installed.

Substrates damaged during paint or transparent finish removal should also be repaired before a new paint or finish is applied.

Materials and Manufacturers. Paint, transparent finish, and associated materials should be as listed in the part of this chapter called "Paint and Transparent Finish Materials," and the discussion earlier about manufacturers also applies here.

In addition to new materials being compatible with each other within each paint or transparent finish system, as is necessary in new buildings, new materials for use in existing buildings must be selected for their compatibility with existing materials that will remain in place. Tests should be conducted in the building to ensure compatibility. Incompatible materials should not be used. Applying water-based paints over solvent-based paints can create serious problems, for example. The water-based paint may lift the oil-based paint from the substrate. All paint applied before 1950 is likely to have an oil base, since water-based paints were unheard of before then.

Some materials are limited in their usability for repainting or refinishing existing painted or finished surfaces. Some transparent finishes containing urethane resins, for example, are not applicable over other finishes and may not even be applicable over a previous coating of the same material without extensive preparation. Lacquer cannot be used over solvent-based paints. The lacquer will act as a paint remover.

Materials should be delivered to the project site in the manufacturer's original packaging. Materials should be packed, stored, and handled carefully to prevent damage.

Paint and Transparent Finish Systems. The discussion of paint and transparent finish systems earlier in this chapter applies to systems to be used in existing buildings, with several differences. The materials themselves are basically the same. Some coats may be eliminated, however, when the existing paint or finish is sound. It is also necessary to ensure that the selected new paint and finish systems are appropriate for repainting or refinishing every existing surface under every condition to be encountered.

Surfaces Usually Painted or Finished. Which existing surfaces are to be repainted or refinished varies greatly from project to project. It may vary from repainting or refinishing every previously painted or finished surface in the building to repainting only those surfaces damaged as a result of other renovation, remodeling, or restoration work being done at the site.

Some possible variations include the following:

Repainting or refinishing, as appropriate, existing interior and exterior previously painted or transparent finished surfaces to remain exposed, whether such surfaces were originally field- or factory-painted or finished, that are scratched, marred, or otherwise damaged during the work being done as part of a remodeling, renovation, or restoration project, regardless of the location of the surfaces.

Repainting or refinishing, as appropriate, every existing interior and exterior previously painted and transparent finished surface to remain exposed, whether they were originally field- or factory-painted or finished, regardless of their location.

Repainting or refinishing, as appropriate, existing interior surfaces when those surfaces are in the same plane with and adjacent to new materials installed as part of a remodeling, renovation, or restoration project. Examples include painting adjacent surfaces as well as new surfaces when a door has been removed and the opening filled; and painting adjacent surfaces when wall material has been removed and a door added.

Repainting or refinishing, as appropriate, existing concealed surfaces that are adjacent to previously painted or transparent finished surfaces when the concealed surfaces are exposed by renovation work.

Painting or finishing, as appropriate, existing items that were not originally painted or transparent finished, but which are of the same type item as new items to be painted or finished. Examples include flashings, piping, ducts, conduit, and the like.

New surfaces which occur as a result of renovation, remodeling, or restoration work should be painted or given a transparent finish in accordance with the same considerations discussed earlier in this chapter for painting and finishing in new buildings.

Surfaces That Are Usually Not Painted or Finished. The discussion earlier in this chapter about new surfaces that are usually not painted or finished applies as well to existing surfaces. Such existing surfaces that were originally painted or finished, however, often are repainted or refinished later. In addition, the following existing surfaces will often not be painted or finished unless they were originally painted or finished.

Existing surfaces that remain concealed and generally inaccessible.

Existing surfaces that are to be covered with new materials, such as gypsum board, glazed wall and ceiling surfacing, ceramic tile, or another new finish.

Existing surfaces that are in satisfactory condition and are left undisturbed during the renovation, remodeling, or restoration project, unless such surfaces are damaged during that work.

Preparation for Painting and Applying Transparent Finishes

General Requirements. Before general painting or finishing begins, a complete wall, space, or item selected for each principal color should be painted or finished for each principal paint and transparent finish that will be used. The sample should show color, finished degree of gloss and texture, materials, and workmanship. The sample should be repainted or refinished until acceptable. The acceptable sample wall, space, or item should then serve as the standard for similar work.

Paint and transparent finish materials should be stored in a single place for each major portion of the work. Storage places should be kept clean. Paint and transparent finish materials should not be stored in closets or other small confined locations.

Oily rags, oil, and solvent-soaked waste should be removed from the buildings at the end of each work day. Precautions should be taken to avoid the danger of fire.

Water-based paints should be applied only when the temperature of the surfaces to be painted and the surrounding air is between 50 and 90 degrees Fahrenheit, unless the paint manufacturer's printed instructions say otherwise.

Solvent-thinned paints and transparent finish materials should be applied only when the temperature of the surfaces to be painted and the surrounding air is between 45 and 95 degrees Fahrenheit, unless the paint or finish manufacturer's printed instructions otherwise recommend.

Paint and transparent finishes should not be applied in snow, rain, fog, or mist; or when the relative humidity exceeds 85 percent; or to damp or wet surfaces; or to extremely hot or cold metal, unless the material manufacturer's printed instructions recommend otherwise. The painting of surfaces exposed to the hot sun should be avoided. Painting may be continued during inclement weather, however, if the areas and surfaces to be painted are enclosed and heated to within the temperature limits specified by the paint manufacturer during the application and drying periods.

Once painting has been started within a building, a temperature of 65 degrees Fahrenheit or higher should be provided in the area where the work is being done. Wide variations in temperatures that might result in condensation on freshly painted surfaces should be avoided.

Inspection and Testing. Areas to be repainted or refinished, and the conditions under which the paint or transparent finish is to be applied, should be inspected carefully. It is not unusual to find conditions other than those that are expected. Unsatisfactory conditions should be corrected before work related to the paint or transparent finish is begun.

Some testing of existing surfaces may be necessary. Some painting contractors are able to conduct such examinations and tests, but many are not. Most national paint manufacturers employ personnel who are trained and equipped to make inspections and tests and to recommend methods for dealing with paint failures.

Tests that may be required include compatibility tests, existing paint thickness tests, and moisture content tests. The results often point directly to the cause of a failure. Test results can also help prevent further failures caused by misdiagnosis or using the wrong methods to make repairs or repaint failed surfaces. Examination may show, for example, whether just a surface coat is peeling or whether the entire thickness of the paint is effected. Microscopic examination may be necessary to determine the number of coats already applied. Chemical testing may be needed to determine the type of paint (oil- or water-based, for example) used in the existing coats.

When existing finishes are to be removed, their removal should be verified. Removal should be complete, including undercoats and other materials that would affect the new paint or finish or show through the new materials. Such bare substrates should be clean and completely free from films or coatings.

Even when surfaces to be painted or finished cannot be put in the proper condition by customary cleaning, sanding, and puttying, proper conditions must be achieved, using extraordinary methods if necessary.

Each coat of a paint or finish system should be inspected and found to be satisfactory before the next coat is applied.

General Requirements for Existing Surface Preparation.   Preparation and cleaning procedures should be done in accordance with the paint or transparent finish material manufacturer's instructions for each specific substrate condition.

Hardware, accessories, machined surfaces, plates, fixtures, and similar items not to be painted or finished should be removed before surface preparation for painting or finishing begins. Alternatively, such items may be protected by surface-applied tape or another type of protection. Even when such items are to be painted, they are often removed to make painting them and the adjacent surfaces easier. After the painting or finishing in each space or area has been completed, the removed items can be reinstalled.

Surface should be cleaned before the first paint or transparent finish coat is applied and, if necessary, also between coats to remove oil, grease, dirt, dust, and other contaminants. Activities should be scheduled so that contaminants will not fall onto wet, newly painted or finished surfaces. Paint or finish should not be applied over dirt, rust, scale, oil, grease, moisture, scuffed surfaces, or other conditions detrimental to formation of a durable paint or finish film.

Mildew should be removed from painted surfaces and neutralized by scrubbing the affected surfaces thoroughly with a solution made by adding two ounces of trisodium phosphate type cleaner and eight ounces of sodium hypochlorite (Chlorox) to one gallon of warm water. A scouring powder may be used, if necessary, to remove mildew spores, but care must be taken to prevent damage to the surface being cleaned. The Brick Institute of America recommends removing mildew from unpainted brick masonry using a mixture of three ounces of trisodium phosphate, one ounce of detergent, eight ounces of sodium hypochlorite, and enough water to make a gallon of solution. Cleaned surfaces should be rinsed with clear water and allowed to dry thoroughly.

Each water-stained interior surface and the interior surface of exterior walls to which paint is to be directly applied should first be primed. The primer should be an oil-based paint product specifically recommended by its manufacturer and the finish coat manufacturer to conceal water stain marks so that they will not show through the finish paint coats. This prime coat should be in addition to the number of coats recommended by the paint manufacturer, as required to produce the paint system to be applied on the affected surface.

Existing surfaces containing natural salts that have not been washed away because rain does not strike the surface should be washed with a solution of trisodium phosphate to remove the salts. Susceptible areas are exterior surfaces under eaves and beneath roof overhangs.

Existing surfaces to be repainted or refinished that are glossy should be roughened by sanding or another appropriate method that will produce a surface tooth sufficient to properly receive the new paint or finish.

Unless there is a compelling reason to do so that is not associated with the paint itself, or the total thickness of the existing paint is 1/16 inch or more, where existing paint is essentially sound, it is necessary to remove only loose and peeling paint, rust, oil, grease, dust, soiling, and other substances that would affect the bond or appearance of the new paint. It is not necessary to remove paint from surfaces that have only minor blemishes, dirt, soot, pollution, cobwebs, insect cocoons, and the like, unless removal is necessary so that repairs can be made to the substrate. Such foreign matter may be removed using a water spray followed by scrubbing with a mild household detergent and water solution. The cleaned surfaces should be rinsed thoroughly with clean water and permitted to dry.

Excess chalking can be removed from existing sound paint using a solution of 1/2 cup household detergent to a gallon of water, and a medium-soft bristle brush. After the chalking has been removed, the surface should be rinsed with a clean water spray and permitted to dry thoroughly. New paint should be applied before chalking starts again.

When paint is stained, the source of the stain should be located and the conditions that produce the stain should be corrected. When the staining is caused by the rusting of metal objects, the metal should be uncovered, hand sanded, and coated with a rust-inhibitive primer. Nails and screws should be countersunk and spot-primed. Holes should be filled with wood filler. Discolorations caused by color extractives in wood substrates should be removed by cleaning with a solution of equal parts denatured alcohol and water. Then the surface should be rinsed, permitted to dry, and covered with a stain-blocking primer formulated for the purpose.

It may be possible to remove existing paint only down to sound paint layers where the existing paint displays crazing, blistering, peeling, or cracking of the top layer or layers only.

It is necessary to remove existing paint down to the bare substrate and correct the underlying problem when the existing paint displays excessive chalking, or blistering, peeling, flaking, cracking, or scaling through the entire thickness (Fig. 5-18). It may also be necessary to remove the existing paint when the introduction of new substrate material prohibits a smooth transition from the existing paint to the new paint. Existing paint that has

Figure 5-18 The paint shown in this photo passed beyond the point of salvation a long time ago. It must be completely removed, the stucco substrate properly prepared, and new paint applied. (*Photo by author.*)

been allowed to build up to more than 1/16 inch in thickness should also be removed down to the bare substrate.

It is necessary to completely remove existing cement-based paints before applying other paint types. New cement-based paint may, however, be used in some paint systems as a primer for other paints. Existing cement-based paint to which a new coat will be applied should be wire-brushed and scrubbed. Mildew, efflorescence, and other contaminants should, of course, be removed.

Where paint has been completely removed, the substrate should be prepared as it would be if it were new.

Paint should be removed by one of the following methods:

Limited paint removal. Where removal is limited to small areas, abrasive methods should be used. Abrasive methods include scraping and sanding. Hand tools should be used in most cases. Gouging of the substrate should be avoided. Mechanical abrasive methods, such as orbital sanders and belt sanders, may be appropriate in some circumstances, but such devices should be used carefully to prevent removing too much material. Rotary sanders, sandblasting, and waterblasting are not appropriate and should not be used for limited paint removal.

Total paint removal using heat. For total paint removal, thermal methods such as an electric heat plate or electric heat gun should be used. Blow torches and other flame-producing methods are dangerous and should not be used for paint removal. Fire insurance requirements should be checked and methods cleared with the insurance provider before any hot-process removal is undertaken.

Total paint removal using chemicals. Chemicals formulated for the purpose, such as solvent-based strippers and caustic strippers, may also be used for total paint removal. Such materials must be used carefully. The manufacturer's directions should be followed exactly. Precautions are necessary to protect workers and others against harm from inhaling vapors, fire, eye damage, chemical poisoning resulting from skin contact with chemicals, and other dangers associated with using chemicals. Lead residue and other harmful substances should be disposed of properly. Chemical strippers must be completely removed before paint is applied. The residue left by some strippers will impede adhesion of the new paint.

Total paint removal using abrasives. Abrasive methods include scraping, sanding, waterblasting, sandblasting, and blasting with other abrasives. One such abrasive product that has been used successfully in some conditions is made from corn cobs. Hand tools should be used where appropriate. Gouging of the substrate should be avoided. Mechanical abrasive methods, such as orbital sanders and belt sanders, may be appropriate in some circumstances, but such devices should be used carefully to prevent removing too much material. Sandblasting is probably not an appropriate means to remove paint from galvanized metal, because preventing damage to the metal while doing it requires more expertise than is commonly available, and because it removes the galvanizing.

Improper removal and disposal of existing chromium- and lead-based paints can create serious health and legal problems. Removal and disposal of such materials is regulated at the national level, and by many state and local jurisdictions as well. For example, material containing more than five parts per million of lead or chromium is classified by the Environmental Protection Agency (EPA) as hazardous waste, and is subject to all applicable regulations. Serious problems can also occur when the materials to be removed contain asbestos, arsenic, or other toxic substances. Even methylene chloride, which is used in many commercial paint removers, is classified as a toxic waste material. Removal and disposal of such materials must be done in strict accordance with laws and recognized safety precautions. Stiff fines and penalties may be imposed on owners, architects, and contractors if such materials are handled or disposed of improperly, regardless of who is at fault. A complete discussion of hazardous materials removal and disposal is beyond the scope of this book. When faced with such a problem, one should obtain copies of the applicable rules and regulations and follow them explicitly. The Steel Structures Painting Council, at the address listed in the Appendix, is a source for current recommendations about cleaning hazardous materials from metals. Their recommendations may also be applicable to other materials, or at least point to other sources. The actual removal of hazardous wastes is best done by professionals who are expert in such work.

Textured surfaces to be painted can usually be prepared using methods similar to those used on any painted surface. Special coatings are not usually painted. Repairs to them are usually made using the same materials as were used in the original application. Care must be exercised when dealing with such products, because many earlier examples contained asbestos. Materials with a factory-applied or integral texture that are to be painted should be cleaned, using methods that will not damage the finishes. Sand paint may be removed entirely by using the same methods used to remove any thick paint coat. Damaged sand-finished paint may be repairable, but most likely at least some of the existing finish will have to be removed first. Repairs can sometimes be made using a compatible paint to which a compatible aggregate, perhaps sand, has been added. Matching such surfaces is a difficult job best left to experienced professionals. Even they should practice producing the damaged finish before attempting to make repairs. After the repairs have been made and are dry, a new coat of paint can be applied to renew the paint.

Transparent finishes may be removed by sanding or scraping, but usually are best removed with chemical removers. On open-grained woods, it may be necessary to use steel wool dipped in the remover to remove the finish from the wood's pores. After the finish has been removed, the remover must be neutralized, using the material recommended by the manufacturer of the remover. It may not be possible, however, to remove an existing penetrating oil stain. Bleaching might remove the stain. If not, the surface must be restained or painted.

Shoulders at the edges of sound paint should be ground smooth and sanded as necessary to remove shoulders, so that flaws do not photograph through the new paint. Where recommended by the paint manufacturer, paint edges may be feathered using drywall joint compound or another method recommended by the paint manufacturer.

Existing surfaces that are to be cleaned and not repainted or refinished should have soiling, stains, dirt, grease, oil, and other unsightly contaminants removed completely. Detergents and cleaning compounds should be compatible with existing surfaces and should not damage existing finishes. Materials and methods to be used should be those recommended by the paint or finish manufacturer. Surfaces damaged due to the use of incorrect products or due to improper use of cleaning agents should be refinished.

Existing paint to be repainted should be scrubbed using a household cleaner and water, rinsed thoroughly with clean water, and allowed to dry before painting.

Water-stained painted surfaces that are to be painted should be coated with a special primer. The primer should be an oil-based coating or paint product specifically recommended by its manufacturer to conceal water stain marks so that they will not show through paint coats. This prime coat should be in addition to the coats recommended by the paint manufacturer for use in the paint system selected.

Before painting or transparent finishing is started in an area, that space should be swept clean with brooms and excessive dust should be removed. After painting or finishing operations have been started in a given area, that area should not be swept with brooms. Necessary cleaning should then be done using commercial vacuum cleaning equipment.

Surfaces to be painted or finished should be kept clean, dry, smooth, and free from dust and foreign matter that would adversely affect adhesion or appearance.

Preparing Concrete, Stone, Concrete Masonry Units, and Other Masonry. Existing unpainted concrete, stone, concrete masonry units, and other masonry to be painted should be prepared to receive paint using exactly the same methods as discussed earlier in this chapter for preparing new surfaces. The previous paragraphs under this heading about existing paint apply when these materials have been previously painted.

Preparing Brick Masonry.   Existing unpainted brick to be painted should be prepared to receive paint by removing efflorescence, chalk, dust, dirt, grease, oils, mildew, mold, moss, and stains. The methods used should not harm the surfaces being cleaned or adjacent surfaces. Acid cleaning solutions should not be used on brick that will be painted.

The alkalinity and moisture content of surfaces to be painted should be ascertained, using appropriate tests. When the surfaces are sufficiently alkaline to blister or burn the paint, the condition must be corrected and the alkalinity neutralized before paint is applied. Zinc chloride and zinc sulfate are often used as a neutralizer. Then an alkali-resistant primer should be used.

The presence of efflorescence, mildew, mold, or moss indicates that there is water present. The underlying condition that is causing the excess moisture to be present must be found and eliminated. Paint should not be applied where the moisture content exceeds that permitted in the paint manufacturer's printed directions or in the Brick Institute of America's *Technical Notes on Brick Construction,* "6: Painting Brick Masonry."

Efflorescence can be removed using clear water and stiff brushes. Moss will succumb to weed killers. Mildew may be removed by steam cleaning, sandblasting, or using a solution of trisodium phosphate, detergent, sodium hypochlorite, and water in the mixture recommended by the Brick Institute of America in *Technical Notes on Brick Construction,* "6: Painting Brick Masonry."

The recommendations of the Brick Institute of America, as contained in the publications listed in "Where to Get More Information" later in this chapter, should be strictly followed when preparing for or painting brick masonry.

Preparing Wood.   Before starting to prepare for repainting or refinishing wood, it should be verified that deteriorated wood has been removed, that wood members to be repaired have been repaired, and that replacement with new wood has been completed. It may be necessary to remove the paint covering in some locations to determine the state of the underlying wood members.

Previously painted or transparent finished surfaces to be repainted or refinished should be cleaned of all dirt, oil, wax, and other foreign substances which might interfere with adhesion or damage the new materials, using scrapers, mineral spirits, and sandpaper, as required. Care should be exercised when using mineral spirits, of course, to ensure that it will not lift the existing paint or finish material. Wood surfaces and edges should be sanded smooth and even before repainting or refinishing is begun and between coats. Dust should be removed after sanding. Residue should be removed from knots, pitch streaks, cracks, open joints, and sappy spots. On wood surfaces to be painted, a thin coat of white shellac should be applied to pitch and resinous sapwood before the prime coat is applied. Small, dry, seasoned knots should be scraped and cleaned and a thin coat of white shellac or other recommended knot sealer should be applied over them before the prime coat is applied. After the primer has been applied, holes and imperfections in the finish surfaces should be filled with putty or plastic wood-filler. The surfaces should be sandpapered smooth when they have dried.

When the surfaces of existing transparent finished wood to be refinished are in good condition, old wax and polish should be removed by wiping with mineral spirits. Rags should be changed frequently. Surfaces should be sanded lightly, until they are dull. Dust should be removed using a tack rag.

When the surfaces of existing transparent finished wood are porous or in poor condition, or are severely discolored or stained (Figs. 5-19 and 5-20), and where the existing finish is scratched or otherwise damaged, the old finish should be removed completely. The surface should then be cleaned and prepared as discussed earlier in this chapter for new wood. Extraordinary cleaning and bleaching may be necessary to remove some stains and discolorations. Application of a new wood stain may be necessary to achieve an acceptable appearance.

Nails fastening wood to be repainted or refinished should be set and such screws should be countersunk where such was not previously done. After the prime coat has been applied and has dried, the nail and screw holes and cracks, open joints, and other defects should be filled with putty or wood filler. Where a transparent finish will be used, the filler should be tinted to match the color of the wood.

Wood that is weathered and displays an open fuzzy grain usually will not hold paint. It must be sanded smooth and coated first. One possible coating mixture contains two parts turpentine and one part linseed oil, but severely weathered wood may require a one to one mix. One coat may be enough, but as many as three are sometimes necessary. After the coating has thoroughly dried, the surfaces should be primed using an oil-alkyd primer.

Preparing Metals.   Existing ferrous metal surfaces, which are not galvanized, shop-coated, or previously painted, and which are to be painted, should be cleaned of oil, grease, dirt, loose mill scale, and other foreign substances

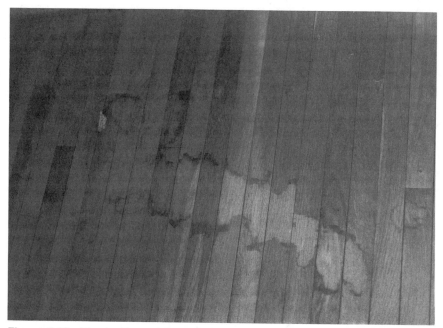

Figure 5-19 The hardwood floor in this photo was stained by a spilled alcoholic beverage which was not immediately cleaned up. It needs refinishing. (*Photo by author.*)

using solvent or mechanical cleaning methods, as recommended by the paint manufacturer. The recommendations of the Steel Structures Painting Council should be followed, unless they conflict with the paint manufacturer's recommendations. Bare and sandblasted or pickled ferrous metal should be given a metal treatment wash before the primer is applied.

It is necessary to remove dirt, oil, grease, and defective paint from shop-coated or previously painted ferrous, galvanized, and nonferrous metal surfaces, using a combination of scraping, sanding, wire brushing, sandblasting, and a cleaning agent recommended by the paint manufacturer. Surfaces should be wiped clean before they are painted. Shoulders should be sanded, if necessary, to prevent telegraphing. Where necessary to produce a smooth finish, existing paint should be stripped to bare metal.

Removal of rust before painting or repainting is essential. Various methods may be used, depending on the amount of rust present and the severity of the problem. Methods include sandblasting, power tool cleaning, hand cleaning, and phosphoric acid wash cleaning. In each case, the recommendations of the Steel Structures Painting Council should be followed.

Successfully painting existing unpainted galvanized metal is one of the

Figure 5-20 The hardwood floor shown in this photo was severely stained by an apartment dweller's pets. It requires complete refinishing. (*Photo by author.*)

more difficult problems related to paints. One reason is that authoritative sources disagree about the proper methods and materials that should be used to prepare galvanized metals to receive paint. Most sources do agree, however, that existing galvanized surfaces will contain white rust. When the galvanizing is damaged, galvanized metal may also contain common rust. Both types of rust, and any other contaminants that may be present, must be removed before paint is applied. Methods for removing rust include wirebrushing, sanding, and blasting. The brush-off blast cleaning method recommended by the Steel Structures Painting Council is often used.

Existing galvanized surfaces may also collect oils and grease; dirt and dust; smoke particles, soot, and other pollutants; and other contaminants, and may still have, on the surface, the residue of oils, waxes, silicons, or silicates applied during the fabrication process. Such materials must be completely removed before paint is applied. Some sources say that galvanized surfaces should be cleaned using mineral spirits or xylol. Others say to use a solvent wash, and specifically recommend against using mineral spirits. Some say to use petroleum spirits. Others say not to use petroleum-based solvents.

After all oils, white rust, and other contaminants have been removed, some sources say that the bare galvanized surfaces should be pretreated. Some sources say to use a diluted vinegar solution, which is weak acetic acid. Others say to use a proprietary acid-bound resinous or crystalline zinc phosphate preparation, or phosphoric acid. Still other sources say that vinegar and other acids should not be used on galvanized metal because they remove some of the galvanizing and leave a residue that can cause an applied paint to peel.

Before a new primer paint is applied, damage to the galvanizing should be repaired using galvanized repair paint that is manufactured for that purpose.

The best solution to determining the proper preparation methods to use on existing unpainted galvanized metals is to ask the paint manufacturer for recommendations, and follow them. At least then there will be someone to complain to if paint failure occurs.

Bare new aluminum should be washed with mineral spirits or turpentine, and allowed to weather one month or be roughened with steel wool.

Bare existing aluminum should be washed with mineral spirits, turpentine, or lacquer thinner as appropriate. It should then be steam-cleaned and wire-brushed to remove loose coatings, oil, dirt, grease, and other substances that would reduce a bond or harm new paint, and to produce a surface suitable to receive new paint.

Bare stainless steel should be washed with a solvent and prepared according to the paint material manufacturer's recommendations.

Bare copper should have stains and mill scale removed by sanding. Then dirt, grease, and oil should be removed using mineral spirits. If a white alkyd primer was used, the copper should be wiped clean using a clean rag which has been wetted with Varsol.

Bare lead-coated copper should be sanded to bright metal. Grease and oil should be removed using a solvent recommended by the metal manufacturer.

Bare terne-coated metal should be cleaned with mineral spirits and wiped dry with clean cloths. Immediately after cleaning, the surfaces should be given a coat of a pretreatment specifically formulated for use on terne-coated metal. The pretreatments should contain 95 percent linseed oil. The pretreatment should be allowed to dry for at least 72 hours, and longer, if necessary to ensure that it is thoroughly dry, before the primer is applied.

Just as welds, abrasions, factory- or shop-applied prime coats, and shop-painted surfaces in new materials should be touched-up after erection and before caulking and application of the first field coat of paint, previously painted surfaces should also be touched-up. On new surfaces, the touch-up material should be the same as the primer. On previously painted surfaces, the touch-up paint should be the same material as the first field coat.

Surfaces which are to be covered should be touched-up before they are concealed. Touch-ups should be sanded smooth.

Preparing Gypsum Board, Plaster, Stucco, and Mineral-Fiber-Reinforced Cement Panels. Existing painted surfaces should be cleaned and prepared, as discussed earlier in this part of this chapter. When an existing paint has been completely removed from gypsum board, stucco, plaster, or mineral-fiber-reinforced cement panels, the bare surfaces should be cleaned and prepared for painting exactly as described earlier in this chapter for new surfaces.

The assumption should be made that older installations of mineral-fiber-reinforced cement panels are probably cement-asbestos panels. Such panels should be handled carefully to prevent danger from the asbestos. Panels should not be broken, sanded, or otherwise handled in such a way as to produce dust or airborne particles, except by specialists trained to handle hazardous substances.

Preparing Other Existing Surfaces. Surfaces not mentioned in this chapter should be prepared for repainting or refinishing in accordance with the paint or finishing material manufacturer's recommendations.

Paint and Transparent Finish Applications

General Application Requirements. The type of material for each coat, location, and use should be as designated for the purpose by the paint or finishing material manufacturer, as stated in manufacturer's published specifications.

Painting and finishing materials should be mixed, thinned, and applied in accordance with manufacturer's latest published directions. Materials should not be thinned unless the manufacturer specifically recommends. Applicators and techniques should be those best suited for the substrate and the type of material being applied.

Sealers and undercoats should not be varied from those recommended by the paint or finish manufacturer.

Materials not in actual use should be stored in tightly covered containers. Containers used in the storage, mixing, and application of paint should be maintained in a clean condition, free of foreign materials and residue.

Materials should be stirred before application to produce a mixture of uniform density, and stirred as required during application. Surface films should not be stirred into the material. Film should be removed and, if necessary, the material strained before it is used.

The extent of painting and finishing on existing, previously painted surfaces may be all surfaces or only part of the surfaces depending on the requirements of the project. When only partial repainting is required, it is usually best to extend the repainting to natural breaks, such as corners or projections.

Existing transparent finishes are seldom only partly refinished, however. Usually, such surfaces are completely refinished. Where a surface cannot be refinished without affecting edges and projections, the edges and projections are usually also refinished.

Repainting and refinishing of existing surfaces should not be started until patchwork, extensions, repair work, and new work in the space have been completed (Fig. 5-21).

When a flat finish is applied over a previously painted surface, it is usually best to apply all of the coats recommended by the paint manufacturer for the selected paint system for new work.

When a semigloss or gloss finish is applied over a previously painted surface, the first coat (primer) should be applied in accordance with the

Figure 5-21 It would be fruitless even to begin preparation for repainting the room shown in this photo before the missing window had been replaced. (*Photo by Stewart Bros., courtesy of Mid-City Financial Corporation.*)

paint manufacturer's specifications, and the final coats should be applied the same as they would be on a new surface.

Before applying new paint or transparent finish coats over existing paint or finishes, it is best to apply the paint or finish in small inconspicuous locations representative of each condition that will be encountered to test for compatibility.

Workmanship should be of a high standard. Painting and finishing should be done by skilled mechanics using proper types and sizes of brushes, roller covers, and spray equipment.

Equipment should be kept clean and in proper condition. The same rollers or brushes used to paint concrete or masonry should not be used on smooth surfaces. Rollers used for a gloss finish and the corresponding primer, should have a short nap.

Materials should be applied evenly and uniformly under adequate illumination. Surfaces should be completely covered and should be smooth and free from runs, sags, holidays, clogging, and excessive flooding. Completed surfaces should be free of brush marks, bubbles, dust, excessive roller stipple, and other imperfections. Where spraying is required or permitted, the paint should be free of streaking, lapping, and pile-up.

The number of coats recommended by the manufacturer for each particular combination of use, substrate, and paint or transparent finish system should be the minimum number used. Primers may be omitted from previously painted surfaces where the existing paint is sound, but all other coats recommended by the manufacturer should be applied to every surface, including previously painted or finished surfaces, even when the existing paint or transparent finish is sound, unless the paint or finish manufacturer specifically recommends otherwise for the particular situation. When properly applied, that number of coats should produce a fully covered, workmanlike, presentable job. Of course, each coat must be applied in heavy body, without improper thinning. When stain, dirt, or undercoats show through the final coat of paint, the defects should be corrected and the surface covered with additional coats until the paint film is of uniform finish, color, appearance, and coverage.

Paint and transparent finish materials should be applied at such a rate of coverage that the fully dried film thickness for each coat will not be less than that recommended by the manufacturer. Special attention should be given to insure that edges, corners, crevices, welds, and exposed fasteners receive a dry film thickness equivalent to that of flat surfaces.

Paint coats should be brushed or rolled onto concrete, concrete masonry units, plaster, and gypsum board. Paint coats on doors and frames and on metal should be applied by brush.

It is necessary to sand lightly between each succeeding enamel or varnish coat. High-gloss paint should be sanded between coats using very fine grit sandpaper. Dust should be removed after each sanding to produce a smooth, even finish.

The tops, bottoms, and side edges of exterior doors should be finished the same as the exterior faces of the doors.

Where a transparent finish other than an oil finish will be applied on the interior of a building over an open-grained wood, such as elm, oak, hickory, or walnut, from which the original finish has been completely removed, paste wood-filler should be applied and wiped across the grain as it begins to flatten. A circular motion should be used to secure a smooth, filled, clean surface with filler remaining in the wood's open grain. Where the wood's stain is to be renewed, the filler is often slightly tinted with the same stain. The filler may also be slightly stained in order to emphasize the grain, where the overall surface will not be stained. After the filler has dried overnight, the surface should be sanded until smooth before the next coat is applied. Steel wool should not be used. Where the next coat is a stain, the filler should be allowed to dry for at least 24 hours before the stain is applied.

Surfaces to be restained and receive a transparent finish should be covered entirely with a uniform coat of stain, using either a brush or clean cloth. Excess stain should then be wiped off.

When a classic oil finish is desired, boiled linseed oil should be thinned with turpentine. The mixture should be between equal parts of oil and turpentine to twice as much oil as turpentine. The actual mixture depends solely on the personal preferences of the applicator. An oil finish may be applied over a stain or directly to bare wood. The oil-turpentine mixture should be brushed liberally onto the wood until no more is absorbed. Then the excess should be wiped off with clean cloths and the surface polished to a sheen. Sufficient pressure should be exerted to melt the oil. This can be achieved with an orbital sander or by hand pressure. After a 48-hour drying period, the entire process should be repeated at least five times.

At plaster to be repainted, adjacent trim and accessories should also be properly prepared and painted.

The first coat of paint or transparent finish material should be applied to surfaces that have been prepared for repainting or refinishing as soon as practicable after preparation and before subsequent surface deterioration. Sufficient time should be allowed between successive coatings to permit proper drying. A minimum of 24 hours is required between interior coats and 48 hours between exterior coats. Surfaces should be recoated until the previous coat has dried to where it feels firm, does not deform, or feel sticky under moderate thumb pressure, and application of another coat of paint or transparent finish materials does not cause lifting or loss of adhesion of the undercoat.

The edges of paint adjoining other materials or colors should be made sharp and clean in straight lines, without overlapping.

At the completion of painting and finishing, damaged paint and finishes should be examined, touched-up and restored where damaged, and left in proper condition.

Clean-up and Protection.   At the end of each work day, while painting and finishing are being done, discarded paint materials, rubbish, cans, and rags should be removed from the site.

When painting and finishing have been completed, glass and other paint-spattered surfaces should be cleaned. Spattered paint should be removed using scraping or other methods that will not scatch or otherwise damage the finished surfaces.

Where to Get More Information

The books, periodicals, and other publications and standards listed here are helpful when selecting or specifying paint and transparent finishes on new and existing surfaces. Some are also helpful in determining how to maintain and repair painted and transparent finished surfaces.

The November 1988 edition of AIA Service Corporation's *Masterspec* "Section 09900 Painting" includes a discussion and detailed listing of the national and regional paint manufacturers operating at that time. The regional manufacturers are broken down into those in the northeast region, southern states and gulf coasts, north central states, southwest region, and northwest states. The author assumes that *Masterspec's* later editions will contain similar data.

"Section 09900 Painting" also includes a discussion of requirements for selecting paint products and descriptions of the components of paints and transparent coating materials that are more extensive than those in this chapter. The author recommends that anyone involved in selecting paint systems or transparent finish materials obtain and refer to a copy.

The editions listed in the Bibliography of U.S. Department of the Navy guide specification "NFGS-09910C, Painting of Buildings (Field Painting)," U.S. Department of the Army "CEGS-09910, Painting, General," and U.S. General Services Administration "PBS: 3-0990, Painting and Finishing, Renovation, Repair, and Improvement" contain excellent guidance for dealing with conditions where new paint is to be applied over an existing surface. Later editions of the Navy and Army guides should be equally helpful. Current editions of the GSA guides, however, are based on *Masterspec* and may not be as specifically helpful for dealing with existing surfaces.

The U.S. General Services Administration Specifications Unit's Federal Specifications applicable to paint and transparent finishes are sometimes out of date or not available and sometimes the quality level they require is substandard. Unfortunately, however, they are often the only applicable standard. Of the many applicable Federal Specifications, some are more often referred to than others. Some of the more-often-referenced ones are included in the following list. The list is by no means complete. Complying with the listed Federal Specification does not necessarily guarantee a satisfactory product for any particular condition.

Federal Specification TT-E-487E, Enamel, Floor and Deck

Federal Specification TT-E-489G, Enamel, Alkyd, Gloss (For Exterior and Interior Surfaces)

Federal Specification TT-E-505A, Enamel, Odorless, Alkyd, Interior, High Gloss, White and Light Tints

Federal Specification TT-E-506K, Enamel, Alkyd, Gloss, Tints and White (For Exterior and Interior Surfaces)

Federal Specification TT-E-508C, Enamel, Interior, Semigloss, Tints and White

Federal Specification TT-E-509B, Enamel, Odorless, Alkyd, Interior, Semigloss, White and Tints

Federal Specification TT-E-527C, Enamel, Lusterless

Federal Specification TT-E-543A, Enamel, Interior, Undercoat, Tints and White

Federal Specification TT-F-322D, Filler Two-component Type: For Dents, Cracks, Small-holes, and Blow Holes. Materials meeting this standard are used as a filler and for patching cracks and seams in wood, metal, concrete, and cement mortar.

Federal Specification TT-F-336E, Filler, Wood, Paste

Federal Specification TT-F-340C, Filler, Wood, Plastic

Federal Specification TT-L-58E, Lacquer, Spraying, Clear and Pigmented for Interior Use

Federal Specification TT-L-190D, Linseed Oil, Boiled (For Use in Organic Coatings)

Federal Specification TT-L-201A, Linseed Oil, Heat Polymerized

Federal Specification TT-P-19C, Paint, Acrylic Emulsion: Exterior

Federal Specification TT-P-25E, Primer Coating, Exterior (Undercoat for Wood, Ready-mixed, White and Tints)

Federal Specification TT-P-29J, Paint, Latex Base, Interior, Flat, White and Tints

Federal Specification TT-P-30E, Paint, Alkyd, Odorless, Interior, Flat White and Tints

Federal Specification TT-P-37D, Paint, Alkyd Resin; Exterior Trim, Deep Colors

Federal Specification TT-P-52D, Paint, Oil (Alkyd Oil) Wood, Shakes and Rough Siding

Federal Specification TT-P-55B, Paint, Polyvinyl Acetate Emulsion, Exterior

Federal Specification TT-P-81E, Paint, Oil, Alkyd, Ready Mixed Exterior, Medium Shades

Federal Specification TT-P-86G, Paint, Red-lead-base, Ready-mixed

Federal Specification TT-P-636D, Primer Coating, Alkyd, Wood and Ferrous Metal

Federal Specification TT-P-641G, Primer Coating; Zinc Dust—Zinc Oxide (For Galvanized Surfaces)

Federal Specification TT-P-645A, Primer, Paint, Zinc Chromate, Alkyd Type

Federal Specification TT-P-650C, Primer Coating, Latex Base, Interior, White (For Gypsum Wallboard)

Federal Specification TT-P-664C, Primer Coating, Synthetic, Rust-inhibiting, Lacquer-resisting

Federal Specification TT-P-791A, Putty, Pure-linseed-oil (For) Wood-sash-glazing

Federal Specification TT-S-176E, Sealer, Surface, Varnish Type, Floor, Wood and Cork

Federal Specification TT-S-300A, Shellac, Cut

Federal Specification TT-S-708A, Stain, Oil; Semi-transparent, Wood, Exterior

Federal Specification TT-S-708A, Stain, Oil Type, Wood, Interior

Federal Specification TT-T-291F, Thinner, Paint, Mineral Spirits, Regular and Odorless

Federal Specification TT-V-86C, Varnish, Oil, Rubbing (For Metal and Wood Furniture)

There are more than 600 ASTM standards related to paint and transparent finish materials. Most of them, however, are testing requirements for the chemicals that might be used in those products and are, therefore, not too helpful in selecting new paint materials or determining the type of paint that might be found in an existing building. The following list contains some ASTM standards that might be helpful.

Standard D 16, Standard Definitions of Terms Relating to Paint, Varnish, Lacquer, and Related Products

Standard D 2833, Standard Index of Methods for Testing Architectural Paints and Coatings

Standard D 3276, Standard Guide for Painting Inspectors (Metal Substrates)

Standard D 3927, Standard Guide for State and Institutional Purchasing of Paint

The following publications of the Steel Structures Painting Council should be available to everyone who is responsible for painting steel.

The 1983 *Steel Structures Painting Manual, Vol. 1, Good Paint Practice, Second Edition.*

The 1983 *Steel Structures Painting Manual, Vol. 2, Systems and Specifications, Second Edition.*

The June 1988 article, "Direct to Rust Coatings," in *American Paint Contractor*, contains a good description of the various materials available for painting rusted metals and their application.

The following Brick Institute of America's *Technical Notes on Brick Construction* contain BIA's recommendations related to painting brick masonry.

The May 1972 edition of technical note "6, Painting Brick Masonry" contains BIA's recommendations for paints and painting methods.

The September/October 1977 edition of technical note "20, Cleaning Brick Masonry" contains BIA's recommendations for cleaning new and existing brick masonry, including, but not limited to, removal of efflorescence and other stains.

The December 1969 edition of technical notes "23, Efflorescence, Causes" and the January 1970 edition of technical note "23A, Efflorescence, Prevention and Control" contain information about the causes of different types of efflorescence and BIA's recommendations for preventing and controlling it.

Anyone involved in designing, maintaining, or repairing painted brick masonry or who is comtemplating painting unpainted brick masonry should obtain and read all four of the listed BIA *Technical Notes.*

The Gypsum Association's 1985 publication *Using Gypsum Board for Walls and Ceilings (GA-201-85)* contains the Gypsum Association's recommendations for painting gypsum board. Anyone responsible for designing, maintaining, or repairing painted gypsum board should read a copy.

The following National Concrete Masonry Association publications contain NCMA's recommendations for painting concrete unit masonry and for maintenance of such painted surfaces. Anyone dealing with designing, maintaining, or repairing painted concrete masonry, or who is contemplating painting unpainted concrete masonry, should obtain and read a copy of each.

1981 NCMA-TEK 10A—Decorative Waterproofing of Concrete Masonry Walls. Of the three publications listed, this is the most definitive. It includes a listing of paint types and NCMA's recommendations regarding each type.

1972 NCMA-TEK 44—Maintenance of Concrete Masonry Walls.

1973 NCMA-TEK 55—Waterproof Coatings for Concrete Masonry.

The National Decorating Products Association's 1988 *Paint Problem Solver* is a definitive source that specifically addresses paint problems and their solutions. It addresses such adhesion problems as alligatoring, blistering, checking, cracking, flaking, excessive chalking, intercoat peeling under eaves and covered porches (scaling or cornflaking); peeling from asbestos shingles, galvanized metal, hardboard siding, concrete, brick and other masonry, plaster, metal doors and garages, wood windows, doors, frames, and other wood; sagging, wrinkling, peeling of latex topcoat from previously painted hard, slick surfaces, and more. It also addresses application problems, including the applicator not holding enough paint, brush marks, cratering, excessive shedding of bristles onto the painted surface, and excessive splatter from rollers. Discolorations are covered, including cedar stain (tannic acid bleed), fading, mildew, rusted nail heads, staining from flashings, and wax bleeding on hardboard siding. Other problems are covered as well, including lap marks, poor hiding, uneven gloss, and painting over wallpaper and other flexible coatings. Everyone responsible for paint maintenance and painting over existing materials should have a copy. It is also available from the Painting and Decorating Contractors of America.

The following Painting and Decorating Contractors of America (PDCA) publications are of varying value, as noted, to those having to deal with the maintenance of paint or painting of existing surfaces. Some critics have rightly pointed out that much of the data in the PDCA publications are also available from some major national paint manufacturers. The advantage of having the PDCA data is twofold. First, it is a source of needed data that is always available, even when the selected manufacturer does not publish similar recommendations, which is often the case. Second, it serves as a second opinion when compared with data printed in other sources.

The 1975 edition of *Painting and Decorating Craftsman's Manual and Textbook, Fifth Edition.* This book is used in training craftspersons.

The 1982 *Painting and Decorating Encyclopedia.* This is a good resource for designers, decorators, and craftspersons. It is not specific to dealing with existing conditions.

The 1986 *Architectural Specification Manual, Painting, Repainting, Wallcovering and Gypsum Wallboard Finishing, Third Edition.* This book was actually published by Specifications Services, Washington State Council of PDCA in Kent, Washington. It should be on the shelf of every person who is responsible for designing, maintaining, and repainting painted surfaces. It includes an evaluation of finishing systems for paint and wallcovering. It also includes discussion of new surface preparation; existing surface preparation; general information and finish schedules for interior and exterior paint finishes; and guide specifications. The guide specifications are not as definitive as some others that are available, and are not based on the current CSI format. The parts on repainting include a discussion of how to handle existing sound, slightly deteriorated, and severely deteriorated paint. Preparation methods are included for wood, metal, concrete, unit masonry, stucco, brick, unglazed tile, concrete floors, plaster, gypsum board, wallcovering, ferrous metal, galvanized metal, copper, and aluminum. It includes a discussion of the cleaning and removal of existing paint by hand, solvent, steam, power tool, burn-off, chemicals, and sandblasting. It also includes removal methods for efflorescence and mildew, and acid etching methods. Other subjects include methods for handling extractive bleeding from cedar, redwood, mahogany, and Douglas fir.

The 1988 *The Master Painters Glossary.* This is an excellent glossary of paint industry terminology that is well worth having.

The following Portland Cement Association (PCA) publications should be available to everyone who is responsible for painting, maintaining, or repainting concrete.

The 1965 "Bulletin D89: Moisture Migration—Concrete Slab-On-Ground Construction."

The 1977 publication "Painting Concrete, IS134T." This publication discusses surface preparation and paint selection and application.

The 1982 publication "Removing Stains and Cleaning Concrete Surfaces."

The 1986 publication "Effects of Substances on Concrete and Guide to Protective Treatments."

C.R. Bennett's 1987 "Paints and Coatings: Getting beneath the Surface" is as complete a general introduction to paints and coatings as one would expect to find in a magazine article. Mike Bauer, in his 1987 letter to the editor, takes issue with some of Mr. Bennett's statements, however. So both pieces should be read together.

The *Consumer Reports* March 1985 article "Concrete Floor Paints" is an excellent discussion of the problems involved in painting concrete floors. It also recommends methods that should be used in preparing to paint and painting concrete floors and compares the various available paint products that will do that job.

Larry Jones's 1984 article "Painting Galvanized Metal" is a good discussion of the problems associated with painting galvanized metal, which is one of the more difficult surfaces to paint successfully. Some of the recommendations in the article are not universally accepted, however.

Bill Kneemiller's 1982 article "Using Exterior Textured Coatings," though somewhat dated now, contains a good introductory look at such materials. The materials discussed there are not truly paint, however, and so are not discussed at length in this book.

Clem Labine's 1982 article "Restoring Clear Finishes" is a good discussion of methods for restoring an existing transparent finish without complete stripping.

Clem Labine's 1984 article "Paint Encrusted Plaster Woes" is a good discussion of the problems associated with stripping paint from surfaces containing small details such as might be found on old plaster moldings.

The National Paint, Varnish and Lacquer Association's (now the National Paint and Coatings Association) publication, *Finishing Hardwood Floors*, discusses both new transparent floor finishes and refinishing.

Patricia Poore's 1981 article "Picking a Floor Finish" discusses maintaining transparent finished floors and the repairing of minor damage such as scuff marks, cigarette burns, and stains. It also compares several different types of clear and stained transparent finishes, including varnish, shellac, and polyurethane.

Patricia Poore's 1985 article "Stripping Paint from Exterior Wood" is an excellent discussion of the subject.

The Old-House Journal's 1982 article "Stripping Paint" is an excellent discussion of the available methods for completely removing existing paint.

The Old-House Journal's 1983 article "48 Paint Stripping Tips" contains practical tips for removing existing paints. Most of the data there is most useful to those actually doing the stripping, however, and is not too helpful in determining whether to strip existing paint or the general method to be used.

The following United States Gypsum Company publications contain troubleshooting recommendations that are generally accepted in the industry as accurate. They should be in the library of everyone who is responsible for maintaining painted plaster or gypsum board surfaces or for repainting such surfaces.

The 1972 publication *Red Book: Lathing and Plastering Handbook, 28th Edition*

The 1987 publication *Gypsum Construction Handbook, Third Edition.*

Kay D. Weeks's and David W. Look's 1982 publication "Exterior Paint Problems on Historic Woodwork" is an excellent discussion of exterior paint failures that is applicable to all conditions, not just historic preservation projects. It gives causes, recommended treatments, methods of paint removal, and some general recommendations for selecting paint for repainting exterior wood.

In addition, the following articles and books may prove of some value:

Able Banov's 1973 book *Paints and Coatings Handbook for Contractors, Architects, and Builders.*

Architectural Technology's 1986 article "Technical Tips: Paints and Coatings Primer."

David R. Black's 1987 article "Dealing with Peeling Paint."

The *Canadian Heritage's* 1985 article "Take It All Off? Advice on When and How to Strip Interior Paintwork."

Sarah B. Chase's 1984 article "Home Work: The ABC's of House Painting."

The Construction Specifications Institute's 1988 Monograph 07M411, Precoated Metal Building Panels, and the 1988 *Specguide* 09900, Painting.

Dan Elswick's 1987 article "Preparing Historic Woodwork for Repainting, Part 2—Thermal and Chemical Cleaning."

The following Gypsum Association publications:

1985, *Recommended Specifications for Application and Finishing of Gypsum Board (GA-216-85).*

1985, *Using Gypsum Board for Walls and Ceilings (GA-201-85).*

1986, *Recommendations for Covering Existing Interior Walls and Ceilings with Gypsum Board.*

Caleb Hornbostel's 1978 book *Construction Materials, Types, Uses, and Applications.*

Larry Jones's 1984 article "Don't Overlook the Heat Plate."

Walter Jowers' 1986 article "Textured Plaster Finishes."

Mary Kincaid's 1982 article "What Paint Experts Say."

Robert Lowes's May/June 1988 article "Abrasive Blasting" in *PWC Magazine.*

Ambrose F. Moormann's 1982–83 article series "Paint and the Prudent Specifier."

Dave Mahowald's 1988 article "Specifying Paint Coatings for Harsh Environments."

Charles R. Martens's 1974 book *Technology of Paints and Lacquers.*

Bill O'Donnell's 1985 article "Reconditioning Floors" and his 1986 article "Unwanted Textured Finish: How to Get Rid of It."

The Old-House Journal's 1983 articles "Our Opinion of 'Peel Away,' " "Paint On Paint," and "Calcimine Concerns;" 1987 article "Exterior Painting: Problems and Solutions;" and 1988 article "Commercial Paint Stripping."

The Reader's Digest *Complete Do-it-yourself Manual.*

The U.S. Department of the Army's 1980 *Painting: New Construction and Maintenance (EM 1110-2-3400).*

The U.S. Department of the Army's 1981 *Corps of Engineers Guide Specifications, Military Construction,* "CEGS-09910, Painting, General."

The U.S. Departments of the Army, the Navy, and the Air Force's 1969 *Technical Manual TM 5-618, Paints and Protective Coatings.*

The U.S. Department of Commerce, National Bureau of Standards 1968 *Organic Coatings BSS 7.*

The Guy E. Weismantel edited 1981 *Paint Handbook.*

Most of the representatives of organizations and associations that the author contacted while gathering the material in this book were quite helpful and cooperative. The sole exception was The National Paint and Coatings Association, where no representative would consent to be interviewed, and the person who answered the telephone would give out no information about the organization. As strange as that attitude may seem at first glance, it may not be strange at all.

It has long been common knowledge in the industry that some paint manufacturers, having recognized that architects do not determine whose paint will be used in most projects, do not identify architects as marketing targets, and do not go out of their way to respond to architect's questions. Even some of the most cooperative of the major national paint manufacturers either do not have knowledgeable architectural representatives in their regional offices, or assign them such large areas to cover that architects have difficulty getting information from them. When the painting project is a new building, the manufacturer's published data will usually suffice. As the number of repainting projects grows, however, architects and other designers need to know more and more about dealing with existing surfaces and failed finishes, and how to solve the many problems involved. For understandable, though lamentable, reasons paint manufacturers and their associations have not always made answers easy to find. That may be one reason so much of the available information referenced in this chapter was produced by the government, by organizations representing contractors, by organizations representing architects, and by the producers of the surfaces to which paint is applied.

Some manufacturers do publish helpful data, of course. They also answer architects' questions, even though often only over the telephone from their home offices. Architects know which manufacturers are the most helpful and respond accordingly. But even those organizations' knowledgeable representatives are sometimes hard to get to. For example, several paint manufacturers' representatives provided data for this chapter. One interview, with a knowledgeable representative of a major national paint manufacturer, took more than six weeks to schedule and was canceled and rescheduled in the process. That representative spent a good deal of time with the author and the meeting yielded much helpful information. But real-project problems, especially those involving failures, must frequently be resolved quickly.

Building owners and painting contractors, because they do decide whose paint will be used, sometimes find manufacturers' representatives more accessible than do architects. When problems arise, an architect is sometimes better off asking one of them to call the paint manufacturer instead of having the call come from the architect, unless, of course, the architect has been able to establish a close relationship with that particular representative.

APPENDIX

Data Sources

NOTE: The following list includes sources of data referenced in the text, included in the Bibliography, or both. **HP** following a source indicates that that source also contains data of interest to those concerned with historic preservation.

Advisory Council on Historic
Preservation
1100 Pennsylvania Avenue, Suite 809
Washington, DC 20004
(202) 786-0503 **HP**

AIA Service Corporation
The American Institute of Architects
1735 New York Avenue, N.W.
Washington, DC 20006
(202) 626-7300

Aluminum Association (AA)
818 Connecticut Avenue N.W.
Washington, DC 20006
(202) 862-5100

American Architectural Manufacturer's Association (AAMA)
2700 River Road, Suite 118
Des Plaines, IL 60018
(312) 699-7310

American Association of State and
Local History
172 Second Avenue, North, Suite 102
Nashville, TN 37201
(615) 255-2971 **HP**

American Concrete Institute
P.O. Box 19150, Redford Station
2400 West Seven Mile Road
Detroit, MI 48219
(313) 532-2600

The American Institute of Architects
1735 New York Avenue, N.W.
Washington, DC 20006
(202) 626-7300

American Institute of Architects
Committee on Historic
Resources
1735 New York Avenue, N.W.
Washington, DC 20006
(202) 626-7300 **HP**

American Institute of Timber
Construction
333 West Hampden Avenue
Englewood, CO 80110
(303) 761-3212

American Iron and Steel Institute
1133 15th Street, Suite 300
Washington, DC 20005
(202) 452-7100

American National Standards Institute (ANSI)
1430 Broadway
New York, NY 10018
(212) 354-3300

American Painting Contractor
American Paint Journal Company
2911 Washington Avenue
St. Louis, MO 63103

American Plywood Association (APA)
P.O. Box 11700
Tacoma, WA 98411
(206) 565-6600

American Society of Heating, Refrigerating and Air-Conditioning Engineers, Inc.
1791 Tullie Circle, N.E.
Atlanta, GA 20329
(404) 636-8400

American Wood Preservers Association
P.O. Box 5283
Springfield, VA 21666
(703) 339-6660

American Wood Preservers Bureau
P.O. Box 6058
2772 South Randolf St.
Arlington, VA 22206
(703) 931-8180

Architectural Technology
The American Institute of Architects
1735 New York Avenue, N.W.
Washington, DC 20006
(202) 626-7300

Architecture
The American Institute of Architects
1735 New York Avenue, N.W.
Washington, DC 20006
(202) 626-7300

Association for Preservation Technology
Box 2487 Station D

Ottawa, ONT K1P 5W6, Canada
(613) 238-1972 **HP**

Association of the Wall and Ceiling Industries—International
1600 Cameron Street
Alexandria, VA 20002
(703) 684-2924

ASTM
1916 Race Street
Philadelphia, PA 19103-1187
(215) 299-5585

Brick Institute of America
11490 Commerce Park Drive, Suite 300
Reston, VA 22091
(703) 620-0010

Building Design and Construction
Cahners Plaza
1350 East Touhy Avenue
P.O. Box 5080
Des Plaines, IL 60018
(312) 635-8800

Campbell Center for Historic Preservation Studies
P.O. Box 66
Mount Carroll, IL 61053
(815) 244-1173 **HP**

Ceilings and Interior Systems Contractors Association
1800 Pickwick Avenue
Glenview, IL 60025
(312) 724-7700

Color Association of the United States
343 Lexington Avenue
New York, NY 10016
(212) 683-9531

Commercial Standard (U.S. Department of Commerce)
Government Printing Office
Washington, DC 20402
(202) 377-2000

Construction Research Council
1800 M Street, N.W., Suite 1040
Washington, DC 20036
(202) 785-3378

The Construction Specifier
601 Madison Street
Alexandria, VA 22314-1791
(703) 684-0300

Decorating Products World
Elkin Mittelman (publishers)
7335 Topango Canyon Boulevard
Canoga Park, CA 91303
(818) 710-1066

Environmental Protection Agency
401 M Street, S.W.
Washington, DC 20460
(202) 829-3535

Exteriors
1 East 1st Street
Duluth, MN 55802
(218) 723-9200

Factory Mutual System
1151 Boston-Providence Turnpike
Norwood, MA 02062
(617) 762-4300

Federal Housing Administration
(U.S. Department of Housing and
    Urban Development)
451 7th Street, S.W.
Washington, DC 20201
(202) 755-5995

Federal Specification (General Ser-
    vices Administration)
Specifications Unit (WFSIS)
7th and D Streets, S.W.
Washington, DC 20406
(202) 472-2205

Forest Products Laboratory
U.S. Department of Agriculture
Gifford Pinchot Drive
P.O. Box 5130
Madison, WI 53705
(698) 264-5600

General Services Administration
General Services Building
18th and F Streets, N.W.
Washington, DC 20405
(202) 655-4000

Gypsum Association
1603 Orrington Avenue
Evanston, IL 60201
(312) 491-1744

Hartford Architecture Conservancy
130 Washington Street
Hartford, CT 06106
(203) 525-0279 **HP**

Heritage Canada Foundation
Box 1358 Station B
Ottawa, ONT K1P 5R4, Canada
(613) 237-1867 **HP**

Historic Preservation
(*See* National Trust for Historic Pres-
    ervation) **HP**

Illinois Historic Preservation Agency
Division of Preservation Services
Old State Capitol
Springfield, IL 62701
(217) 782-4836 **HP**

Institute for Applied Technology/
Center for Building Technology
National Bureau of Standards
U.S. Department of Commerce
Washington, DC 20540
(202) 342-2241

*Journal of Special Coatings and
    Linings*
(*Contact* Steel Structures Painting
    Council)

Library of Congress
1st Street, N.E.
Washington, DC 20540
(202) 287-5000 **HP**

McGraw-Hill Book Company
1221 Avenue of the Americas

New York, NY 10020
(212) 997-2271

National Alliance of Preservation
    Commissions
Hall of the States
444 North Capitol Street, N.W.,
    Suite 332
Washington, DC 20001
(202) 624-5490 **HP**

National Association of Architectural
    Metal Manufacturers
600 South Federal Street, Suite 400
Chicago, IL 60605
(312) 922-6222

National Association of Corrosion
    Engineers
1440 South Creek
Houston, TX 77084
(713) 492-0535

National Bureau of Standards (NBS)
(*See* National Institute of Standards
    and Technology)

National Concrete Masonry
    Association
P.O. Box 781
Herndon, VA 22070
(703) 435-4900

National Decorating Products
    Association
1050 North Lindbergh Boulevard
St. Louis, MO 63132
(314) 991-3470

National Fire Protection Association
Batterymarch Park
Quincy, MA 02269
(617) 770-3000

National Forest Products Association
1619 Massachusetts Avenue, N.W.
Washington, DC 20036
(202) 797-5800

National Institute of Standards and
    Technology (formerly National
    Bureau of Standards)
Gaithersburg, MD 20234
(301) 975-2000

National Institute of Standards and
    Technology (formerly National
    Bureau of Standards)
Center for Building Technology
Gaithersburg, MD 20234
(301) 975-5900

National Paint and Coatings
    Association
1500 Rhode Island Avenue, N.W.
Washington, DC 20005
(202) 462-6272

National Paint, Varnish and Lacquer
    Association
(*See* National Paint and Coatings
    Association)

National Preservation Institute
P.O. Box 1702
Alexandria, VA 22313
(703) 393-0038 **HP**

National Trust for Historic
    Preservation
1785 Massachusetts Avenue, N.W.
Washington, DC 20036
(202) 673-4000 **HP**

Occupational Safety and Health Ad-
    ministration (U.S. Department of
    Labor)
Government Printing Office
Washington, DC 20402
(202) 783-3238

The Old-House Journal
69A Seventh Avenue
Brooklyn, NY 11217
(718) 636-4514

Painting and Decorating Contractors
    of America
3913 Old Lee Highway, Suite 33B

Fairfax, VA 22030
(703) 359-0826

Painting and Decorating Contractors
of America
Washington State Council
27606 Pacific Highway South
Kent, WA 98032
(206) 941-8823

Portland Cement Association
5420 Old Orchard Road
Skokie, IL 60077
(312) 966-6200

The Preservation Press
National Trust for Historic
Preservation
1785 Massachusetts Avenue, N.W.
Washington, DC 20036
(202) 673-4000 **HP**

Preservation Resource Group
5619 Southampton Drive
Springfield, VA 22151
(703) 323-1407 **HP**

Product Standards of NBS (U.S. Department of Commerce)
Government Printing Office
Washington, DC 20402
(202) 783-3238

*PWC Magazine: Painting and Wall-
covering Contractor* (Magazine
of Painting and Decorating Contractors of America)
130 West Lockwood Street
St. Louis, MO 63119
(314) 961-6644

Resilient Floor Covering Institute
966 Hungerford Drive, Suite 12-B
Rockville, MD 20850
(301) 340-8580

Rubber Manufacturer's Association
1400 K Street, N.W.
Washington, DC 20005
(202) 682-1338

Sheet Metal and Air Conditioning
Contractors National Associa-
tion, Inc.
8224 Old Courthouse Road
Vienna, VA 22180
(703) 790-9890
For publications contact:
SMACNA Publications Department
P.O. Box 70
Merrifield, VA 22116
(703) 790-9890

Society for the Preservation of New
England Antiquities
141 Cambridge Street
Boston, MA 02114
(617) 227-3956 **HP**

Southern Pine Inspection Bureau
4709 Scenic Highway
Pensacola, FL 32504
(904) 434-2611

Steel Structures Painting Council
4400 Fifth Avenue
Pittsburgh, PA 15213
(412) 578-3327

Technical Preservation Services
U.S. Department of the Interior,
Preservation Assistance Division
National Park Service
P.O. Box 37127
Washington, DC 20013-7127 **HP**
(202) 343-7394

Truss Plate Institute
583 D'Onofrio Drive, Suite 200
Madison, WI 53719
(608) 833-5900

Underwriters Laboratories, Inc.
333 Pfingsten Road
Northbrook, IL 60062
(312) 272-8800

U.S. Department of Agriculture
14th Street and Independence Avenue, S.W.

Washington, DC 20250
(202) 447-4929

U.S. Department of Commerce
14th Street and Constitution Avenue, N.W.
Washington, DC 20230
(202) 377-2000

U.S. Department of Commerce for
PS (Product Standard of NBS)
Government Printing Office
Washington, DC 20402
(202) 783-3238

U.S. Department of the Interior
National Park Service
P.O. Box 37127
Washington, DC 20013-7127
(202) 343-7394 **HP**

U.S. General Services Administration
Historic Preservation Office
18th and F Streets, N.W.
Washington, DC 20405
(202) 655-4000 **HP**

U.S. General Services Administration
Specifications Unit
7th and D Streets, S.W.

Washington, DC 20407
(202) 472-2205/2140

United States Gypsum Company
101 South Wacker Drive
Chicago, IL 60606
(312) 606-4000

Van Nostrand Reinhold
115 Fifth Avenue
New York, NY 10003
(212) 254-3232

West Coast Lumber Inspection
Bureau
P.O. Box 23145
Portland, OR 97223
(503) 639-0651

Western Wood Products Association
1500 Yeon Building
Portland, OR 97204
(503) 224-3930

John Wiley and Sons
605 Third Avenue
New York, NY 10158
(212) 850-6000

Bibliography

NOTE: Most items in the Bibliography are followed by one or more numbers in brackets. Those numbers list the chapters in this book to which that bibliographical entry applies.

The **HP** following some entries in the Bibliography indicates that that entry has particular significance for historic preservation projects.

Sources for many of the entries, including addresses and telephone numbers, are listed in the Appendix.

Acoustical Materials Association. AMA 1-II Ceiling Sound Transmission Test by the Two Room Method. Glenview, IL: Ceilings and Interior Systems Contractors Association. [3]

AIA Service Corporation. *Masterspec*, Basic: Section 06100, Rough Carpentry, 8/86 Edition. The American Institute of Architects. [2, 3]

————. *Masterspec*, Basic: Section 09510, Acoustical Ceilings, 5/86 Edition. The American Institute of Architects. [3]

————. *Masterspec*, Basic: Section 09521, Acoustical Wall Panels, 8/86 Edition. The American Institute of Architects. [3]

————. *Masterspec*, Basic: Section 09650, Resilient Flooring, 8/84 Edition. The American Institute of Architects. [4]

————. *Masterspec*, Basic: Section 09800, Special Coatings, 2/88 Edition. The American Institute of Architects. [5]

————. *Masterspec*, Basic: Section 09900, Painting, 2/84 Edition. The American Institute of Architects. [5]

————. *Masterspec*, Basic: Section 13070, Integrated Ceilings, 8/83 Edition. The American Institute of Architects. [3]

Allen, Edward. 1985. *Fundamentals of Building Construction: Materials and Methods.*

New York: Wiley. [2, 3, 4, 5]

American Concrete Institute. 1980. ACI 302.1R-80: Guide for Concrete Floor and Slab Construction. American Concrete Institute. [2, 4]

American Institute of Timber Construction. *Timber Construction Standards*. Englewood, CO: American Institute of Timber Construction. [2]

————. *Timber Construction Manual*. Englewood, CO: American Institute of Timber Construction. [2]

American Paint Contractor. 1988. Direct to Rust Coatings. *American Paint Contractor*, June, 65(6):8–19. [5]

American Plywood Association. 1987. APA Source List: Plywood Underlayment for Use under Resilient Finish Flooring. American Plywood Association. [4]

————. 1988. APA Data File: Installation and Preparation of Plywood Underlayment for Thin Resilient (Non-textile) Flooring. American Plywood Association. [4]

American Wood Preservers Association. *Book of Standards*. Springfield, VA: American Wood Preservers Association. [2, 3]

Architectural Technology. 1986. Technical Tips: Paints and Coatings Primer. *Architectural Technology*, July/August: 64–65. [5]

Association for Preservation Technology. 1969. *Bulletins of APT, Volume 1*. Ottawa, Ontario: APT. **HP**

ASTM. Standard A 123, Specifications for Zinc (Hot-Galvanized) Coatings on Products Fabricated from Rolled, Pressed, and Forged Steel Shapes, Plates, Bars, and Strips. ASTM. [5]

————. Standard A 446, Specifications for Sheet Steel, Zinc-Coated (Galvanized) by the Hot-Dip Process, Structural (Physical) Quality. ASTM. [5]

————. Standard A 525, Specifications for General Requirements for Steel Sheet Zinc-Coated (Galvanized) by the Hot-Dip Process. ASTM. [5]

————. Standard C 635, Metal Suspension Systems for Acoustical Tile and Lay-In Panel Ceilings. ASTM. [3]

————. Standard C 636, Installation of Metal Ceiling Suspension Systems for Acoustical Tile and Lay-In Panels. ASTM. [3]

————. Standard C 919, Practices for Use of Sealants in Acoustical Applications. ASTM. [3]

————. Standard D 16, Standard Definitions of Terms Relating to Paint, Varnish, Lacquer, and Related Products. ASTM. [5]

————. Standard D 1779, Specification for Adhesive for Acoustical Materials. ASTM. [3]

————. Standard D 2833, Standard Index of Methods for Testing Architectural Paints and Coatings. ASTM. [5]

————. Standard D 3276, Standard Guide for Painting Inspectors (Metal Substrates). ASTM. [5]

————. Standard D 3927, Standard Guide for State and Institutional Purchasing of Paint. ASTM. [5]

————. Standard E 84, Test Method for Surface Burning Characteristics of Building Materials. ASTM. [3, 4]

————. Standard E 119, Fire Tests for Building Construction and Materials. ASTM. [3]

————. Standard E 580, Recommended Practice for Application of Ceiling Suspension Systems for Acoustical Tile and Lay-In Panels in Areas Requiring Seismic Restraint. ASTM. [3]

————. Standard E 648, Test Method for Critical Radiant Flux of Floor Covering Systems Using a Radiant Heat Energy Source. ASTM. [4]

————. Standard E 662, Test Method for Specific Optical Density of Smoke Generated by Solid Materials. ASTM. [4]

Banov, Able. 1973. *Paints and Coatings Handbook for Contractors, Architects, and Builders*. Farmington, MI: Structures Publishing Co. [5]

Bartlett, Thomas L. 1983. Cleaning with Corncobs. *The Construction Specifier*, February, 36(2):6–7. [5]

Batcheler, Penelope Hartshorne. 1968. Paint Color Research and Restoration. Nashville, TN: American Association for State and Local History, Technical Leaflet 15. **HP** [5]

Bauer, Mike. 1987. Letterbox: Too Much to the Imagination? *The Construction Specifier*, July, 40(7):7–8. [5]

Bennett, C. R. 1987. Paints and Coatings: Getting Beneath the Surface. *The Construction Specifier*, February, 40(2):36–41. [5]

Black, David R. 1987. Dealing with Peeling Paint. *North Carolina Preservation*, Dec/Jan/Feb, 66:16–17. [5]

Blatterman, Joan F. 1988. Details Underfoot. *Architectural Record*, July: 118–121. [4]

Bodner, James. 1988. From the Bottom Up: Good Beginnings for Floorcovering Installations. *The Construction Specifier*, April, 41(4):98–105. [4]

Bower, Norman F. 1985. Insurance by the Gallon. *The Construction Specifier*, April, 38(4):96–99. [5]

Brick Institute of America. 1972 (May). *Technical Notes on Brick Construction*, 6, Painting Brick Masonry. Reston, VA: Brick Institute of America. [5]

————. 1977 (Sept/Oct). *Technical Notes on Brick Construction*, 20, Cleaning Brick Masonry. Reston, VA: Brick Institute of America. [2, 5]

————. 1969 (Dec). *Technical Notes on Brick Construction*, 23, Efflorescence, Causes. Reston, VA: Brick Institute of America. [2, 5]

————. 1970 (Jan). *Technical Notes on Brick Construction*, 23A, Efflorescence, Prevention and Control. Reston, VA: Brick Institute of America. [2, 5]

Brunnell, Gene. 1977. *Built to Last: A Handbook on Recycling Old Buildings*. Washington, DC: The Preservation Press. **HP**

Canadian Heritage. 1985. Take It All Off? Advice on When and How to Strip Interior Paintwork. *Canadian Heritage*, Dec/Jan:44–47. [5]

Cattell, D. 1988. *Specialist Floor Finishes: Design and Installation*. London: Blackie and Sons Ltd. [4, 5]

Ceilings and Interior Systems Contractors Association. *Acoustical Ceilings: Use and Practice*. Glenview, IL: Ceilings and Interior Systems Contractors Association. [3]

Chase, Sarah B. 1984. Home Work: The ABC's of House Painting. *Historic Preservation*, August, 36(4):12–14. [5]

City Limits. 1986. Exterior Paints. *City Limits*, October, 11(8):15. [5]

Commercial Renovation. 1987. The 1987 Premier Renovation Architects. *Commercial Renovation*, December, 9(6):24–44. **HP**

———. 1987. Product Focus: Ceiling Systems Reach New Heights in Performance and Design. *Commercial Renovation*, December, 9(6):48–53. [3]

Construction Specifications Institute. 1988. CSI Monograph Series, 07M411, Precoated Metal Building Panels. Alexandria, VA: The Construction Specifications Institute. [5]

———. 1988. *Specguide* 09900, Painting. Alexandria, VA: The Construction Specifications Institute. [5]

Construction Specifications Institute, Construction Specifications Canada. 1983. *Masterformat*. Alexandria, VA: The Construction Specifications Institute.

Construction Specifier, The. 1985. Painting to Protect. *The Construction Specifier*, April, 38(4):92–94,123. [5]

———. 1987. A Resource Guide to the 3Rs: Restoration, Renovation and Rehabilitation. *The Construction Specifier*, July, 40(7):102–114. [1, 2, 3, 4, 5]

Consumer Reports. 1985. Concrete-Floor Paints. *Consumer Reports*, March, 50(3), 166–69. [5]

Diereck, X. M. 1975. Metal Ceilings in the U.S. *Association for Preservation Technology Bulletin* (Ottawa, Canada) 7(2):83–98. **HP** [3]

Elswick, Dan. 1987. Preparing Historic Woodwork for Repainting, Part 2—Thermal and Chemical Cleaning. *The New South Carolina State Gazette*, Spring, 19(3):4–5. [5]

Factory Mutual Systems. *Approval Guide*. Norwood, MA: Factory Mutual Systems. [3, 4, 5]

———. *Loss Prevention Data Sheets*. Norwood, MA: Factory Mutual Systems. [3, 4, 5]

Fisher, Thomas. 1983. When the Rains Come. *Architecture*, July. [5]

Fossel, Peter V. 1985. Refinish an Old Floor. *Americana*, Sept/Oct, 13(4):85–88. [5]

Foster, Al. 1988. Textured Coatings: Acoustical in the Name Doesn't Necessarily Mean the Product Absorbs Noise. *PWC Magazine*, May/June, 50(3):34–40. [5]

Freund, Eric C., and Gary L. Olsen. 1985. Renovating Commercial Structures: A Primer. *The Construction Specifier*, July, 38(7):36–47. [2]

Gypsum Association. 1985. *Recommended Specifications for Application and Finishing of Gypsum Board (GA-216-85)*. Evanston, IL: Gypsum Association. [5]

———. 1985. *Using Gypsum Board for Walls and Ceilings (GA-201-85)*. Evanston, IL: Gypsum Association. [5]

———. 1986. *Recommendations for Covering Existing Interior Walls and Ceilings with Gypsum Board (GA-650-86)*. Evanston, IL: Gypsum Association. [5]

Hardingham, David. 1980. Preparing for Painting. *The Old-House Journal*, October:133–36. [5]

Harris, D. A. 1983. Ceilings for the Office of the Future. *The Construction Specifier*, July, 36(7):58–65. [3]

Harsfield, S. W. 1982. A Valid Question. *The Construction Specifier*, December, 35(12):7. [5]

Harvey, John. 1972. *Conservation of Buildings*. London: Baker. **HP**

Herman, Frederick. 1981. Refinishing Floors: Think Twice Before Sanding. *The Old-House Journal*, Feb., 27,44–45. [5]

Historic Preservation. 1983. Home Work. *Historic Preservation*, July/August, 35(4):10–11. **HP** [5]

———. 1984. Home Work: Spring Fixes. *Historic Preservation*, April, 36(2):14–17. **HP** [5]

Hornbostel, Caleb. 1978. *Construction Materials, Types, Uses, and Applications*. New York: Wiley. [5]

Hughes, Richard L., and Martin F. Johnson. 1983. Acoustic Design in Natatoriums. *The Construction Specifier*, July, 36(7):67–70. [3]

Insall, Donald W. 1972. *The Care of Old Buildings Today: A Practical Guide*. London: The Architectural Press. **HP**

Jones, Larry. 1984. Painting Galvanized Metal. *The Old-House Journal*, Jan–Feb, 12(1):10–11. [5]

———. 1984. Don't Overlook the Heat Plate. *The Old-House Journal*, Jan–Feb, 12(1):12–13. [5]

Jowers, Walter. 1986. Textured plaster finishes. *The Old-House Journal*, March, 14(2):75–77. [5]

Kincaid, Mary. 1982. What Paint Experts Say. *The Construction Specifier*, July, 35(5):54–61. [5]

———. 1983. Making Open Office Acoustics Work. *The Construction Specifier*, July, 36(7):70–76. [3]

Kneemiller, Bill. 1982. Using Exterior Textured Coatings. *The Construction Specifier*, July, 35(5):63–68. [5]

Kosmatka, Steven H. 1987. Floor-Covering Materials and Moisture in Concrete. *The Construction Specifier*, May, 40(5):35–36. [4]

Labine, Clem. 1982. Restoring Clear Finishes. *The Old-House Journal*, November, 10(11):221,238–41. [5]

————. 1984. Paint Encrusted Plaster Woes. *The Old-House Journal*, July, 12(6):111–24. [5]

Lawrence, Brenda J. 1983. Battling Mildew in Virginia. *The Old-House Journal*, March, 11(2):37–39. [5]

Lowes, Robert. 1988. Abrasive Blasting. *PWC Magazine*, May/June, 50(3):26–30. [5]

Maberry, Gregg A. 1987. Acoustical Wall Panels: The Space Silencer. *The Construction Specifier*, February, 40(2):50–52. [3]

Mahowald, Dave. 1988. Specifying Paint Coatings for Harsh Environments. *The Construction Specifier*, October, 41(10):13–16. [5, 6]

Martens, Charles R., The Sherwin-Williams Company. 1974. *Technology of Paints and Lacquers*. New York: Robert E. Krieger Publishing Company. [5]

Maruca, Mary. 1984. 10 Most Common Restoration Blunders. *Historic Preservation*, October, 36(5):13–17. **HP**

Matero, Frank G. 1987. Investigating Historic Architectural Paints: An Introduction. *The Construction Specifier*, July, 40(7):66–68. [5]

Mazzur, Richard P. 1986. Resilient Flooring: Choose Carefully. *The Construction Specifier*, March, 39(3):70–75. [4]

McDonald, Timothy B. 1987. Technical Tips: Coatings that Protect against the Corrosion of Steel. *Architecture*, July:101–2. [5]

Moit, Dan. 1988. Coatings for Metals: Preplan the Selection. *Metal Architecture*, September, 4(9):8. [5]

Moormann, Ambrose F., Jr. 1982. Paint and the Prudent Specifier. *The Construction Specifier*, July, 35(5):69–71. [5]

————. 1982. Paint and the Prudent Specifier: If You Must Paint Concrete Floors. *The Construction Specifier*, September, 35(9):60–62. [5]

————. 1983. Paint and the Prudent Specifier: Working Hard to Look Natural. *The Construction Specifier*, January, 36(1):84–87. [5]

Moreno, Elena Marcheso. 1987. Failures Short of Collapse. *Architecture*, July: 91–94. [2]

Munger, Charles G. 1984. *Corrosion Prevention by Protective Coatings*. Houston, TX: National Association of Corrosion Engineers. [5]

National Concrete Masonry Association. 1972 NCMA-TEK 44—Maintenance of Concrete Masonry Walls. National Concrete Masonry Association. [2, 5]

————. 1973 NCMA-TEK 55—Waterproof Coatings for Concrete Masonry. National Concrete Masonry Association. [5]

————. 1981. NCMA-TEK 10A—Decorative Waterproofing of Concrete Masonry Walls. National Concrete Masonry Association. [5]

National Decorating Products Association. 1988. *Paint Problem Solver*. St. Louis, MO: National Decorating Products Association. (Also available from the Painting and Decorating Contractors of America.) [5]

National Forest Products Association. *National Design Specifications for Wood Construction.* Washington, DC: National Forest Products Association. [2, 3]

————. *Manual for House Framing.* Washington, DC: National Forest Products Association. [2, 3]

National Paint, Varnish and Lacquer Association. *Finishing Hardwood Floors.* Washington, DC: National Paint, Varnish and Lacquer Association. [5]

National Trust for Historic Preservation. 1985. *All About Old Buildings—The Whole Preservation Catalog.* Washington, DC: The Preservation Press. (This is an extensive reference work containing the names and addresses of many organizations active in the historic preservation field and lists of publications sources. Anyone facing a preservation problem should obtain this catalog as soon as possible. It will save much time in finding the right organization or data source.) HP

Naval Facilities Engineering Command. 1983 (March). Guide Specifications Section 06100, Rough Carpentry. Department of the Navy. [2]

————. 1983 (May). Guide Specifications Section 09500, Acoustical Treatment. Department of the Navy. [3]

————. 1982 (November). Guide Specifications Section 09650, Resilient Flooring. Department of the Navy. [4]

————. 1986 (February). Guide Specifications Section 09910C, Painting of Buildings (Field Painting). Department of the Navy. [5]

O'Donnell, Bill. 1985. Reconditioning floors. *The Old-House Journal,* December, 13(10):201,218–19. [5]

————. 1986. Unwanted Textured Finish: How to Get Rid of It. *The Old-House Journal,* October:374–77. [5]

Old-House Journal, The. 1982. Stripping Paint. *The Old-House Journal,* December: 249–52. [5]

————. 1983. 48 Paint Stripping Tips. *The Old-House Journal,* March, 11(2):44–45. [5]

————. 1983. Ask OHJ: Restoring Graining. *The Old-House Journal,* March, 11(2):49. [5]

————. 1983. Ask OHJ: The Gilt Complex. *The Old-House Journal,* March, 11(2):49. [5]

————. 1983. Our Opinion of 'Peel Away.' *The Old-House Journal,* May, 11(4):80. [5]

————. 1983. Ask OHJ: Paint On Paint. *The Old-House Journal,* November, 11(9):202. [5]

————. 1983. Ask OHJ: Calcimine Concerns. *The Old-House Journal,* November, 11(9):202. [5]

————. 1984. Tips On Mixing Paint. *The Old-House Journal,* Jan–Feb, 12(1):8–9. [5]

————. 1985. Reconditioning Floors. *The Old-House Journal,* December, 13(10):203,218–19. [5]

————. 1985. Stripping Clinic. *The Old-House Journal,* December, 13(10):212B. [5]

————. 1987. Exterior Painting: Problems and Solutions. *The Old-House Journal,* Sep/Oct:35–39. [5]

————. 1988. Commercial Paint Stripping. *The Old-House Journal,* July/August, 16(4):29–33. [5]

Olin, Harold Bennet, John L. Schmidt, and Walter H. Lewis. 1983. *Construction Principles, Materials and Methods, Fifth Edition.* Chicago, IL: United States League of Savings Institutions. [2, 4]

O'Neil, Edward F., and James E. McDonald. 1976, February. *An Evaluation of Selected Instruments Used to Measure the Moisture Content of Hardened Concrete, Technical Report C-76-1.* Vicksburg, MS: Concrete Laboratory, U.S. Army Engineer Waterways Experiment Station. [4]

Painting and Decorating Contractors of America. 1975. *Painting and Decorating Craftsman's Manual and Textbook, Fifth Edition.* Falls Church, VA: Painting and Decorating Contractors of America. [5]

————. 1982. *Painting and Decorating Encyclopedia.* Falls Church, VA: Painting and Decorating Contractors of America. [5]

————. 1984. *Painting and Wallcovering: A Century of Excellence.* Falls Church, VA: Painting and Decorating Contractors of America. [5]

————. 1986. *Architectural Specification Manual, Painting, Repainting, Wallcovering and Gypsum Wallboard Finishing, Third Edition.* Kent, WA: Specifications Services, Washington State Council of PDCA. [5]

————. 1988. *The Master Painters Glossary.* Falls Church, VA: Painting and Decorating Contractors of America. [5]

————. 1988. *PDCA Estimating Guide.* Falls Church, VA: Painting and Decorating Contractors of America. [5]

Petersen, Maurice R. 1984. Finishes on Metals: A View from the Field—Part 1. *The Construction Specifier,* 37(12):36–39. [5]

————. 1985. Finishes on Metals: A View from the Field—Part 2. *The Construction Specifier,* 38(1):70–73. [5]

Phillips, Morgan W. 1975. Some Notes on Paint Research and Reproduction. *Association for Preservation Technology, Technical Bulletin* (Ottawa, Canada) 7(4),14–19. **HP** [5]

Poore, Patricia. 1980. Sanding a Parquet Floor. *The Old-House Journal,* November: 168–71. [5]

————. 1981. Picking a Floor Finish. *The Old-House Journal,* May:107–13. [5]

————. 1985. Stripping Paint from Exterior Wood. *The Old-House Journal,* December, 13(10):207–11. [5]

Portland Cement Association. 1965. Bulletin D89: Moisture Migration—Concrete Slab-on-Ground Construction. Portland Cement Association. [2, 4, 5]

———. 1977. Painting Concrete, IS134T. Portland Cement Association. [2, 5]

———. 1982. Removing Stains and Cleaning Concrete Surfaces. Portland Cement Association. [2, 5]

———. 1983. Concrete Floors on Ground. Portland Cement Association. [2, 4, 5]

———. 1986. Effects of Substances on Concrete and Guide to Protective Treatments. Portland Cement Association. [2, 5]

———. 1988. *Design and Control of Concrete Mixtures, Thirteenth Edition.* Skokie, IL: Portland Cement Association. [2, 5, 6]

Progressive Builder. 1987. Finishing Wood Exteriors. *Progressive Builder*, January, 12(1):11–20. [5]

Ramsey/Sleeper, The AIA Committee on Architectural Graphic Standards. 1981. *Architectural Graphic Standards, Seventh Edition.* New York: Wiley. [2, 3, 4, 5]

Reader's Digest Association. 1973. *Reader's Digest Complete Do-it-yourself Manual.* Pleasantville, NY: The Reader's Digest Association. [3, 4, 5]

Resilient Floor Covering Institute. *SV-1, Resilient Floor Covering Institute Recommended Specification for Resilient Floor Covering—Vinyl Plastic Sheet.* Rockville, MD: Resilient Floor Covering Institute. [4]

———. *Recommended Work Procedures for Resilient Floor Coverings.* Rockville, MD: Resilient Floor Covering Institute. [4]

Scharfe, Thomas. 1988. New Metal Coating Technologies Enhance Design Opportunities. *Building Design and Construction*, June, 29(6):86–89. [5]

Sherwin-Williams. 1988. Painting and Coating Systems for Specifiers and Applicators. Sherwin-Williams. [5]

Sivinski, Valerie A. 1986. Preserving Historic Materials: Ferrous Metals. *Architecture*, November:108–9. [5]

Smith, Baird M. 1984. Moisture Problems in Historic Masonry Walls: Diagnosis and Treatment. Technical Preservation Services, U.S. Department of the Interior, Technical Report. **HP**

Southern Pine Inspection Bureau. *Standard Grading Rules for Southern Pine Lumber.* Pensacola, FL: Southern Pine Inspection Bureau. [2, 3]

Staehli, Alfred M. 1985. Historic Preservation: Where to Find the Facts. *The Construction Specifier*, July, 38(7):50–53. **HP**

Stahl, Frederick A. 1984. *A Guide to the Maintenance, Repair, and Alteration of Historic Buildings.* New York: Van Nostrand. **HP**

Stanwood, Les. 1983. Taking a Look at Noise. *The Construction Specifier*, July, 36(7):50–55. [3]

———. 1983. Acid Rain: A Cloudy Issue. *The Construction Specifier*, November, 36(11):74–79. [5]

Steel Structures Painting Council. 1983. *Steel Structures Painting Manual, Vol. 1, Good Paint Practice, Second Edition.* Pittsburgh, PA: Steel Structures Painting Council. [5]

———. 1983. *Steel Structures Painting Manual, Vol. 2, Systems and Specifications, Second Edition.* Pittsburgh, PA: Steel Structures Painting Council. [5]

Stoddard, Brooke C. 1987. Home Work: Picking the Right Paint Color. *Historic Preservation,* September/October, 39(5):16–19. **HP** [5]

Truss Plate Institute. *Design Specifications for Light Metal Plate Connected Wood Trusses.* Madison, WI: Truss Plate Institute. [2]

Underwriters Laboratories. *Building Materials Directory—Class A, B, C: Fire and Wind Related Deck Assemblies.* Northbrook, IL: Underwriters Laboratories, Inc. [3]

———. *Fire Resistance Directory—Time/Temperature Constructions.* Northbrook, IL: Underwriters Laboratories. [3, 4, 5]

United States Gypsum Company. 1972. *Red Book: Lathing and Plastering Handbook, 28th Edition.* Chicago, IL: United States Gypsum Company. [5]

———. 1987. *Gypsum Construction Handbook, Third Edition.* Chicago, IL: United States Gypsum Company. [5]

U.S. Department of Army. *Technical Manual TM 5-801-2 Historic Preservation Maintenance Procedures.* Washington, DC: Department of the Army. **HP**

———. 1977 (amended 1982). Corps of Engineers Guide Specification CEGS-09500, Acoustical Treatment. U.S. Department of the Army. [3]

———. 1980. *Painting: New Construction and Maintenance (EM 1110-2-3400).* Washington, DC: Superintendent of Documents, U.S. Government Printing Office. [5]

———. 1981. Corps of Engineers Guide Specifications, Military Construction, CEGS-09910, Painting, General. Office of the Chief of Engineers, Department of the Army. [5]

U.S. Departments of Army, the Navy, and the Air Force. 1969. *Technical Manual TM 5-618, Paints and Protective Coatings.* Washington, DC: U.S. Government Printing Office. [5]

U.S. Department of Commerce. *PS 1—U.S. Product Standard for Construction and Industrial Plywood.* Washington, DC: U.S. Department of Commerce. [4]

———. *PS 20—American Softwood Lumber Standard.* Washington, DC: U.S. Department of Commerce. [2, 3]

———. 1968. *Organic Coatings BSS 7.* Washington, DC: U.S. Department of Commerce, National Bureau of Standards. [5]

U.S. General Services Administration. 1971 (November). Public Building Services Guide Specification, Section 0951, Acoustical Tile—Adhesive Set. U.S. General Services Administration. [3]

———. 1969 (February). Public Building Services Guide Specification, Section 3-0990, Painting and Finishing, Renovation, Repair and Improvement. U.S. General Services Administration. [5]

———. 1969 (January). Public Building Services Guide Specification, Section 4-0990.01, Painting and Finishing. U.S. General Services Administration. [5]

U.S. General Services Administration, Federal Construction Council. 1972. Federal Construction Guide Specification Section 09500, Acoustical Treatment. U.S. General Services Administration, Federal Construction Council. [3]

————. 1979. Federal Construction Guide Specification Section 09513, Ceiling Lay-In Panels (and Suspension Systems) for Use with Background Masking. U.S. General Services Administration, Federal Construction Council. [3]

U.S. General Services Administration Specifications Unit. Federal Specification L-F-475, Floor Covering, Vinyl, Surface (Tile and Roll), with Backing. Washington, DC: GSA Specifications Unit. [4]

————. Federal Specification L-F-001641, Floor Covering, Translucent or Transparent Vinyl Surface with Backing. Washington, DC: GSA Specifications Unit. [4]

————. Federal Specification RR-T-650, Treads, Metallic and Non-Metallic. Washington, DC: GSA Specifications Unit. [4]

————. Federal Specification SS-S-118: Sound Controlling (Acoustical) Tiles and Panels. Washington, DC: GSA Specifications Unit. [3]

————. Federal Specification SS-W-40, Wall Base: Rubber and Vinyl Plastic. Washington, DC: GSA Specifications Unit. [4]

————. Federal Specification SS-T-312, Tile, Floor: Asphalt, Rubber, Vinyl, Vinyl Composition. Washington, DC: GSA Specifications Unit. [4]

————. Federal Specification TT-E-487E, Enamel, Floor and Deck. [5]

————. Federal Specification TT-E-489G, Enamel, Alkyd, Gloss (For Exterior and Interior Surfaces). [5]

————. Federal Specification TT-E-505A, Enamel, Odorless, Alkyd, Interior, High Gloss, White and Light Tints. [5]

————. Federal Specification TT-E-506K, Enamel, Alkyd, Gloss, Tints and White (For Exterior and Interior Surfaces). [5]

————. Federal Specification TT-E-508C, Enamel, Interior, Semigloss, Tints and White. [5]

————. Federal Specification TT-E-509B, Enamel, Odorless, Alkyd, Interior, Semigloss, White and Tints. [5]

————. Federal Specification TT-E-527C, Enamel, Lusterless. [5]

————. Federal Specification TT-E-543A, Enamel, Interior, Undercoat, Tints and White. [5]

————. Federal Specification TT-F-322D, Filler Two-component Type: For Dents, Cracks, Small-holes, and Blow Holes. [5]

————. Federal Specification TT-F-336E, Filler, Wood, Paste. [5]

————. Federal Specification TT-F-340C, Filler, Wood, Plastic. [5]

————. Federal Specification TT-L-58E, Lacquer, Spraying, Clear and Pigmented for Interior Use. [5]

————. Federal Specification TT-L-190D, Linseed Oil, Boiled (For Use in Organic Coatings). [5]

———. Federal Specification TT-L-201A, Linseed Oil, Heat Polymerized. [5]

———. Federal Specification TT-P-19C, Paint, Acrylic Emulsion: Exterior. [5]

———. Federal Specification TT-P-25E, Primer Coating, Exterior (Undercoat for Wood, Ready-mixed, White and Tints). [5]

———. Federal Specification TT-P-29J, Paint, Latex Base, Interior, Flat, White and Tints. [5]

———. Federal Specification TT-P-30E, Paint, Alkyd, Odorless, Interior, Flat White and Tints. [5]

———. Federal Specification TT-P-37D, Paint, Alkyd Resin; Exterior Trim, Deep Colors. [5]

———. Federal Specification TT-P-52D, Paint, Oil (Alkyd Oil) Wood. [5] Shakes and Rough Siding. [5]

———. Federal Specification TT-P-55B, Paint, Polyvinyl Acetate Emulsion, Exterior. [5]

———. Federal Specification TT-P-81E, Paint, Oil, Alkyd, Ready Mixed Exterior, Medium Shades. [5]

———. Federal Specification TT-P-86G, Paint, Red-lead-base, Ready-mixed. [5]

———. Federal Specification TT-P-636D, Primer Coating, Alkyd, Wood and Ferrous Metal. [5]

———. Federal Specification TT-P-641G, Primer Coating; Zinc Dust—Zinc Oxide (For Galvanized Surfaces). [5]

———. Federal Specification TT-P-645A, Primer, Paint, Zinc Chromate, Alkyd Type. [5]

———. Federal Specification TT-P-650C, Primer Coating, Latex Base, Interior, White (For Gypsum Wallboard). [5]

———. Federal Specification TT-P-664C, Primer Coating, Synthetic, Rust-inhibiting, Lacquer-resisting. [5]

———. Federal Specification TT-P-791A, Putty, Pure-linseed-oil (For) Wood-sash-glazing. [5]

———. Federal Specification TT-S-176E, Sealer, Surface, Varnish Type, Floor, Wood and Cork. [5]

———. Federal Specification TT-S-300A, Shellac, Cut. [5]

———. Federal Specification TT-S-708A, Stain, Oil; Semi-transparent, Wood, Exterior. [5]

———. Federal Specification TT-S-708A, Stain, Oil Type, Wood, Interior. [5]

———. Federal Specification TT-T-291F, Thinner, Paint, Mineral Spirits, Regular and Odorless. [5]

———. Federal Specification TT-V-86C, Varnish, Oil, Rubbing (For Metal and Wood Furniture). [5]

———. Federal Specification ZZ-T-001237, Tread, Stair, Flexible and Semi-rigid Type Rubber and Vinyl. Washington, DC: GSA Specifications Unit. [4]

————. Federal Specification MMM-A-00150: Adhesive for Acoustical Materials. Washington, DC: GSA Specifications Unit. [3]

Weeks, Kay D., and David W. Look. 1982. Exterior Paint Problems on Historic Woodwork. Washington, DC: Technical Preservation Services, U.S. Department of the Interior, Preservation Brief No. 10. **HP** [5]

Weinstein, Nat. 1984. How to Match Paint Colors. *The Old-House Journal,* Jan–Feb, 12(1):7–9. [5]

Weismantel, Guy E. (Ed.) 1981. *Paint Handbook.* New York: McGraw-Hill. [5]

West Coast Lumber Inspection Bureau. *Standard Grading Rules for West Coast Lumber.* Portland, OR: West Coast Lumber Inspection Bureau. [2]

Western Wood Products Association. *Grading Rules for Western Lumber.* Portland, OR: Western Wood Products Association. [2]

————. *Grade Stamp Manual.* Portland, OR: Western Wood Products Association. [2]

————. *A-2, Lumber Specifications Information.* Portland, OR: Western Wood Products Association. [2]

————. *Western Woods Use Book.* Portland, OR: Western Wood Products Association. [2]

————. *Wood Frame Design.* Portland, OR: Western Wood Products Association. [2]

Wilson, Forrest. 1984. *Building Materials Evaluation Handbook.* New York: Van Nostrand. [2, 5]

————. 1985. Building diagnostics. *Architectural Technology,* Winter: 22–41. [2]

Wright, Gordon. 1987. Trends in Specifying Architectural Coatings. *Building Design and Construction,* June, 28(6):188–92. [5]

Index

Accessories, for acoustical ceilings, 47, 50–51
Acoustical baffles. *See* Acoustical wall panels and baffles
Acoustical ceilings, 33–89
  acoustical materials and finishes, 34–39, 49, 82
  cleaning of, 82, 82–85
  definitions, 33–34
  failure in,
    evidence of, 66–70
    reasons for, 58–66
  installing new materials, 51–58
  installing over existing materials, 85–89
  repairing, refinishing, and reinstalling removed materials, 70–85
  support systems, 39–49, 53–56, 70–82, 86–89
    metal furring for, 42–43, 54, 70–74, 77–79, 87–88
    metal suspension systems for, 43–49, 54–56, 70–74, 79–82, 86, 88–89

    structural classification of, 47
    wood furring for, 39–42, 53–54, 70–77, 86–87
Acoustical performance
  of acoustical ceilings, 36–37
  of acoustical wall panels and baffles, 99
  of integrated ceilings, 91
Acoustical wall panels and baffles, 94–130
  acoustical materials, 94–99
  cleaning of, 126
  definitions, 33, 34
  failure in,
    evidence of, 114–17
    reasons for, 109–14
  finishes, 99
  installing new materials, 105–9
  installing over existing materials, 128–30
  repairing, refinishing, and reinstalling removed materials, 117–28
  support systems for, 100–103, 106–7, 109, 117–26, 128–30

Acoustical wall panels *(cont.)*
  baffle hangers, 109, 124–26, 128–29, 130
  metal furring for, 103, 107, 117–21, 123–24, 128–29, 129–30
  wood furring for, 100–103, 106–7, 121–23, 129
Architects
  as help for building owners, 5–6
  as help for other architects and engineers, 11
  how to use this book, xvi
  prework on-site examinations by, 13–14
  professional help for, 11–12
Asphalt tile flooring, 139

Back-mounted acoustical panels, 96, 109
Baffles, acoustical, 97–99, 109
Brick, 207–8, 258
Building owners, *See* Owners, building
Building materials manufacturer. *See* Product manufacturer

Coatings. *See* Paint and transparent finishes
Concrete, 206–7, 257–58
Concrete masonry units, 206–7, 257–58
Concrete structure. *See* Steel and concrete structure
Concrete substrates. *See* Solid substrates
Condensation, as a cause of damage, 29–30
Consultant, specialty, 9–14
Contractors
  general building, xvii–xviii, 6–7, 12–15
  specialty, 7–8, 12
Cork tile flooring, 139–40

Damage consultant. *See* Consultant, specialty

Edging strips, 146
Efflorescence, 25, 28. *See also* Paint and transparent finishes
Emergencies, xvi, 4
Engineers
  as help for building owners, 5–6
  as help for architects and other engineers, 11
  how to use this book, xvi–xvii
  prework on-site examinations by, 13–14
  professional help for, 11–12

Finish, definition, 2
Finish failure, 2–3
Fire resistance
  of acoustical ceilings, 51
  of acoustical wall panels and baffles, 103–5
  of resilient flooring, 133
Forensic consultant. *See* Consultant, specialty

Gypsum board, 211, 263

Integrated ceilings, 89–93
  acoustical materials, 89–90
  air diffusion through, 90
  cleaning of, 93
  definitions, 34
  failure in, 92–93
  installing new materials, 91
  installing over existing materials, 93
  lighting component of, 90–91
  repairing, refinishing, and reinstalling removed materials, 93
  support systems, 90

Leaks, as a cause of damage, 29
Light reflectance coefficient (LR), 35, 36
Linoleum flooring, 142

Masonry, 206–7, 257–58
Masonry substrates. *See* Solid
    substrates
Materials manufacturer. *See* Product
    manufacturer
Metal, 209–11, 259–63
Metal acoustical ceilings. *See*
    Acoustical ceilings
Metal furring, 16–17. *See also*
    Acoustical ceilings; Acoustical
    wall panels and baffles
Metal wall panels, 96–97, 109
Mineral-fiber-reinforced cement
    panels, 211–13, 263

Noise reduction coefficient (NRC),
    35, 36, 37, 60, 99

Owners, building
    how to use this book, xvi
    prework on-site examinations by,
        13
    professional help for, 4–11
    what to do in an emergency, xvi,
        4–5

Paint and transparent finishes, 185–
    276
    cleaning of, 237–47
    colors and finishes, 189–90
    composition of, 191–94
    definitions, 185–87
    failure in,
        evidence of, 229–37
        reasons for, 216–28
    installing new materials, 200–201,
        213–16
    installing over existing materials,
        247–50, 263–67
    manufacturers, 187–88
    materials, 187–99
    preparation of existing substrates,
        251–63
    preparation of new substrates, 203–
        13
    removal of, 255–57

repairing of, 237–47
systems, 194–95
Plaster, 211–13, 263
Product manufacturer, 8–9, 11–12

Reducer strips, 146
Resilient flooring, 132–184
    accessories for, 143–46
    adhesives for, 147, 148–49
    cleaning and polishing materials
        for, 149
    cleaning of,
        new materials, 159–60, 183
        existing materials, 167–170, 175
    definitions, 132
    failure in,
        evidence of, 164–66
        reasons for, 160–64
    finishing of, 159–60
    installing new materials, 149–59
    installing over existing materials,
        175–83
    materials, 133–42, 168–69
    primers for, 148
    repairing, refinishing, and
        protecting existing materials,
        167–75
    selecting materials, 133–36
    substrates for
        preparing existing, 171–72
        preparing new, 151–54
        as they affect material selection,
            134–35
    underlayments for, 147–49
Rubber sheet flooring, 142
Rubber tile flooring, 137–38
Rubber wall base. *See* Resilient
    flooring

Sheet flooring, installing, 155–56
Solid substrates, 24–26. *See also*
    acoustical ceilings; Acoustical
    wall panels and baffles;
    Resilient flooring; Paint and
    transparent finishes
Sound insulation blankets, 109

Sound transmission class (STC), 35, 36, 37, 60
Speech privacy noise isolation class (NIC′), 35, 36, 37
Spline-mounted acoustical panels, 94–96, 107–9
Stains, 196–97
Stair nosings, 143, 145–46
Stair risers, 143–45
Stair stringers, 143–45
Stair treads, 143, 145–46
Steel and concrete structure, 16–24
    failure of, 18
    movement of, 23–24
Stone, 206–7, 257–58
Stucco, 211–13, 263
Substrates. *See* Solid substrates; Supported substrates
Supported substrates, 26–29

Transparent finishes, 197–99
    systems, 199

Vinyl composition tile flooring, 137
Vinyl sheet flooring, 140–42
Vinyl tile flooring, 136–37
Vinyl wall base. *See* Resilient flooring

Wall base, 143–45
Welding rod, 146
Wood, 208–9, 258–59
Wood furring, *See* Acoustical ceilings; Acoustical wall panels and baffles
Wood structure, 16–24
    failure of, 18–23
    movement of, 23–24